Heinemann Experimental Chemistry Series

General Editors:
A. J. Mee
Martin Rogers, *Headmaster, Malvern College*

Advisory Panel:
Professor E. H. Coulson, *Organizer, Nuffield Advanced Chemistry Project*
Professor H. F. Halliwell

3
Chemical Analysis, Chromatography, and Ion Exchange

Students' Text

Note : only those titles marked * have been published at the time of publication of the present volume.

Heinemann Experimental Chemistry Series

3
Chemical Analysis, Chromatography, and Ion Exchange

Students' Text

R. T. Allsop, B.A. (Cantab)
Lecturer in Education, University of Hong Kong

J. A. D. Healey, M.A. (Oxon)
Principal Lecturer in Chemistry, Chester College

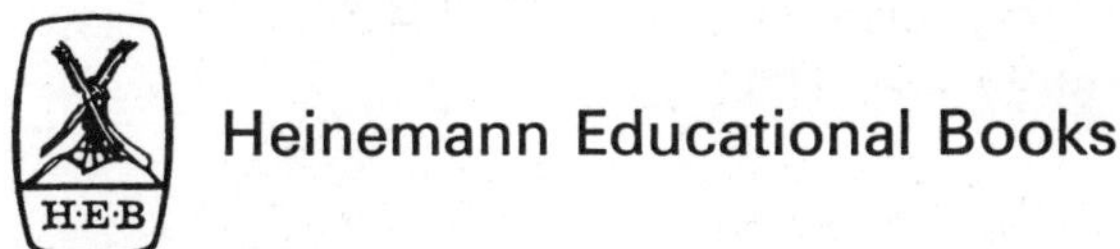

Heinemann Educational Books

Heinemann Educational Books Ltd

London Edinburgh Melbourne Auckland Toronto
Hong Kong Singapore Kuala Lumpur
Ibadan Nairobi Johannesburg
Lusaka New Delhi

ISBN 0 435 65954 5

© R.T. Allsop and J.A.D. Healey 1974

First published 1974

Published by Heinemann Educational Books Ltd
48 Charles Street, London W1X 8AH

Printed in Great Britain by
William Clowes & Sons Limited
London, Beccles and Colchester

Preface to the Series

The richness of the British tradition of science teaching lies in its experimental work. Nowhere in the world has there been for so long such an emphasis on experiments performed both by teachers and students. The books in this series have been written to sustain that tradition and to inject new life into it.

Much of the practical work in chemistry which was done by students in the past consisted merely of the repetition of experiments, the results of which were already known, or in the preparation of compounds by following a rigid set of instructions. In either case there was little room for thought on the part of the student. Modern chemistry teaching seeks to improve on this situation by emphasizing that the aim of practical work is to discover something, not merely to copy what has been done before. Of course, experimental work at this level is not, in fact, research, but there is no reason why it should not be presented as if it were. To the student it *is* something new that he is finding out.

These books have been designed to cover the whole area of advanced school chemistry and introductory courses at colleges and universities. A series of books, each centred on a specified topic, offers considerable advantages over a single volume. It is possible to include more material and to present it in a very flexible way. Teachers can choose those volumes which suit their own particular course and students are not stuck with a heavy volume containing a good deal of irrelevant material. Moreover, each book can be written by a teacher, expert in a particular field.

What is new in this series is the method of treatment. The approach is based on discovery by the students. They are given sufficient information to perform the experiments in an investigational manner and to think about them, but not so much that the performance becomes a meaningless ritual. The experimental work is not an end in itself. The experiments serve some definite purpose in relation to the theory work. By means of questions students are led to examine the results of their work, not only with regard to the degree of accuracy they may have attained but to deduce theoretical points arising from the experiments. In this way the practical work becomes an integral part of the chemistry course and not a mere adjunct to it.

We have attempted to include a number of modern experiments which are new to the school laboratory. Teachers will recognize the many 'classical' experiments which also appear, although these have been treated with a different emphasis.

The books are intended to allow students to proceed at their own pace. This does not mean, of course, that the teacher is absolved from the responsibility of over-seeing the student's work, but it does mean that students can get on without having a lecture period beforehand in which the teacher explains what has to be done in the practical session. It is hoped that the use of the books will foster independence of thought among students and will thus enable the teacher to assume his proper role of counsellor and guide rather than instructor.

For each *Students' Text* a *Teachers' Guide* has been prepared. It gives hints about the conduct of the experiments and the conclusions that can be drawn from them, and contains the answers to most of the questions posed in the *Students' Texts*. The list of materials required for each experiment is provided in the *Students' Texts*.

We cannot hope to have produced a series of books which is completely free from error and we hope that users will inform us of any lapses. It would be foolish to claim, too, that the experiments described are incapable of improvement and we should be glad to hear of any way in which this can be done.

Specific acknowledgment of the considerable amount of help received in the writing of these books will be found in the Introduction to the individual volumes. We are, however, deeply indebted to Prof. E. H. Coulson and Prof. H. F. Halliwell for acting as consultants in the preparation of the series as a whole. They gave very valuable advice from which all the books have benefited.

The study of chemistry through experiments is not only great fun, it is an excellent education. We hope that the teaching of experimental chemistry will be encouraged and enriched by this series.

A. J. Mee
Martin Rogers

1974

Contents

8: Analysis of Brass

9: Gas Analysis

10: The Use of Sodium Thiosulphate in Volumetric Analysis

11: Investigation of Some Polymers

12: Thin-layer and Two-way Chromatography

13: Using the Flow of a Bubble in Solution

Introduction

This book is concerned with chemical analysis. To analyse something is to find out about its composition. This involves questions about what the substance is, what it contains, and how much of each component is in it. When we are dealing with 'how much' the process is called quantitative analysis but if the problem is only 'what', it is called qualitative analysis. Before a quantitative analysis is possible, it is usually necessary to carry out a qualitative analysis, because the techniques used in quantitative analysis will depend on the substances which are present.

Analysis is one of the most frequent activities of a chemist. It is used constantly in industry to test the raw materials received in the factory and to monitor the quality of the products. The raw materials may be as simple as the water used in the steam turbines for producing power, or as complicated as a biochemical substance used in a pharmaceutical product. The substances concerned may not be only the useful ones. Many factories have to carry out continuous analysis of the gaseous effluents which they release from their chimneys or the liquid effluents which they pass into a neighbouring river. Analysis is used in Public Health departments to test the quality of drinking water, the purity of foodstuffs, and so on. In any form of chemical research it is usually necessary to find out by analytical techniques the amount or nature of substances formed and processes carried out. Much modern advance in biological knowledge has involved the precise analysis of the complex materials present in the living cell. The physicist, the metallurgist, and the geologist all require a detailed knowledge of the composition of the materials with which they are dealing. In fact, a comprehensive list of the uses of analysis would be enormous.

An analyst is like a detective. It is not his habit to work totally in the dark. The detective knows what the crime was; he knows the sort of people who in the past have committed similar crimes; he knows where to find these people and he knows which lines of investigation are likely to prove fruitful. If you were to take a material to an analyst and say 'will you analyse this?', his first question would probably be 'what is it?'. This might seem a ridiculous question from the man who is supposed to be giving the answer but what he means is 'where has it come from?' or 'how was it made?' or 'what do you think is wrong with it?'. He must have clues if he is going to start an investigation. If you said to an analyst 'what is dissolved in this water?' he would eventually tell you but he would take a very long time. If you said to him 'I think this water may have come through lead piping. Will it be fit to drink?' he will carry out an analysis of the lead content very quickly. So you will find that all the analyses in this book are in a limited field and it is important that you should always be sure of the particular question you are trying to answer.

Qualitative analysis may involve an enormous number of different techniques but there are common principles included in all of them. Essentially a property is selected for each possible component which is unique to that component amongst all the others which may be present, and which is not interfered with by any of the other components. The property may be a physical one such as solubility or density. Thus you could find out if a sample of sand contained gold by panning it. Gold has a sufficient density to fall to the bottom and remain in the pan, whilst the less dense materials are washed away by the running water. Of course there might be other substances present which have a similar density and possibly they might look like gold. If, however, you are using this technique to analyse for the presence of gold, you are making the assumption that like-natured substances could not be present in the sand.

Very often in an analysis the material undergoes a chemical reaction and it is the properties of the products which are recognized. A mineral which gives a blue solution after it has been roasted and treated with an acid may contain copper. This method might fail if the mineral contained nickel because the green colour of the nickel(II) ion could mask the blue colour of the copper(II) ion. Much of the qualitative analysis in this book involves you setting up your own criteria for the analysis. It could be said that you are building up a key and when you are carrying out the analysis you are finding if the key fits the lock. Very often a textbook on analysis consists of a series of recipes. The book contains the keys and you follow a sequence of tests to identify the locks which they fit. When you are carrying out an analysis, be certain in your own mind whether you are using someone else's key or your own and whether you are within the limitations which restrict the use of the key.

Quantitative analysis also uses a large number of different techniques. Here a property has to be chosen which varies with the composition. The percentage of ethanol in a spirit can be found by determining the density. The proportion of iron in a sample of soil can be found by acting on an acidic extract of the soil with potassium thiocyanate and measuring the intensity of the red colour produced by the iron(III) thiocyanate complex formed. The properties of the components may interfere with one another and it may be necessary to separate the components before the analysis is carried out. The above method for determining iron would fail if insoluble soil particles were not first removed. Frequently the quantitative analysis makes use of a chemical reaction in which the amount of reagents needed to complete the reaction are measured or the amount of the product is determined. This makes several very important assumptions. Consider a method for determining the concentration of sodium carbonate in a solution by reacting it with hydrochloric acid.

$$Na_2CO_3 + 2HCl \longrightarrow 2NaCl + H_2O + CO_2$$

If the method involved measuring the volume of carbon dioxide evolved, we would be assuming that the reaction took place in such a way that precisely one mole of carbon dioxide was produced for every mole of sodium carbonate in the solution, that all the sodium carbonate had undergone reaction, and that all the carbon dioxide produced had been collected and measured. The degree of accuracy of the method may be entirely dependent on the limitations within which these assumptions can be made. This same analysis could involve measuring the amount of hydrochloric acid required and this would mean that the point at which all the sodium carbonate had reacted would have to be detected. If an indicator was used, the assumption would be that the indicator changed colour when precisely two moles of hydrochloric acid had been added to one mole of sodium carbonate. The accuracy of this assumption will depend on a number of factors as you will discover in Chapter 2.

Certain techniques and apparatus are so frequently used in quantitative analysis that special reference must be made to them.

Weighing

Be aware of the accuracy and sensitivity of the balance you are using. You are unlikely to be able to use a top weighing, direct reading balance to a greater accuracy than ± 10 mg. However if you are weighing as little as 1 g, this error will only be ± 1 per cent. Are other errors in your experiment greater than this? If so you need not be concerned about the limitations of your balance. Do you increase or decrease the possible percentage error by weighing out a greater amount of substance?

If you have weighed something accurately, always ensure that you use every bit of it. Considerable losses occur when a material is transferred from one vessel to another and good analytical technique often involves reducing these transference losses to a minimum. It is ridiculous to weigh something to within 10 mg on a piece of filter paper and then simply tip it into the reaction vessel. Use a weighing bottle from which all traces of material can be removed by washing with water.

Making up Standard Solutions

A standard solution is one which contains a known weight of dissolved substance in a given volume of solution. The volume of a solvent may change when a substance is dissolved in it and you cannot make a standard solution by adding a measured volume of solvent to a given weight of substance. A standard flask is used, often called a volumetric flask or a graduated flask. The weighed substance is transferred to the flask and the solvent is added until it reaches a precisely marked level. You will probably use Grade B volumetric flasks which are accurate to about 0.1 per cent.

The following technique will help you to make up accurate standard solutions. Use a weighing bottle to weigh out the solute, first weighing the bottle empty. Transfer the solid to a beaker, washing out the bottle several times with distilled water. Roughly add not more than half the volume of water which you will need to fill the standard flask. Stir the solution until all the solute has dissolved. Be very careful to avoid splashing, and ensure that you do not lose solution by taking out the glass rod and laying it on the bench. If the solute dissolves rather slowly, you may heat the beaker at this stage but you must cool the solution before it is poured into the standard flask. If a standard flask is heated, it will expand and it may not contract to exactly the same volume when it cools again. Hence it will be permanently damaged. When the solute has dissolved, transfer the solution to the standard flask using a glass funnel and pouring down a glass rod. Take the last drop on the rod and put it back in the beaker. Wash the beaker with distilled water and transfer the washings to the flask. Do this at least twice more and finally wash down the funnel and glass rod before they are removed. Make up the solution to the mark on the flask, adding the last portion of distilled water with a teat pipette to avoid overshooting. Remember that all volumetric equipment is calibrated to be read at the bottom of the meniscus. Cork the flask and mix very thoroughly by inverting the flask at least ten times. At this stage you should be able to answer with conviction the question 'has all the solid finished up in this standard solution ?'.

The Pipette

This instrument is designed to deliver a given volume of liquid. A 10 cm^3 Grade B pipette is accurate to ± 0.04 cm^3. A pipette with a chipped tip is not accurate. When the pipette is held vertically, the liquid drains, leaving a precise amount on the walls of the pipette. The instrument is calibrated to account for this but errors are introduced if the pipette is not allowed to drain sufficiently. A rule of thumb is to count a number of seconds equal to the volume of the pipette in cm^3 after the liquid has run out. The small drop left in the tip has also been accounted for in the calibration but the surface tension at the tip affects the volume of this drop. After the pipette has been allowed to drain, the tip should be made to touch the wall of the receiving vessel. This ensures that the correct volume remains in the pipette. On no account should the liquid be blown out from the pipette.

The pipette must obviously be washed out with the liquid which is going to be taken in it. Otherwise you will not know that the given volume of this liquid has been delivered. The same applies to any vessel to which you transfer the liquid before it is pipetted. You should consider very carefully what you are going to use to wash out the vessel into which the liquid is to be run from the pipette. If the liquid is to be used for a titration, the vessel should be washed with distilled water. Can you explain why ?

On no account must concentrated acids or alkalis, or other poisonous substances be sucked into a pipette by mouth. There are suction bulbs made for use in such cases.

The Burette

This instrument is designed to measure the volume of a liquid which has been delivered from it. A 25 cm^3 Grade B burette is accurate to ± 0.08 cm^3. Like a pipette, a burette takes time to drain and you should not read it immediately the tap has been closed. A common error in using a burette is to fail to ensure that the tip below the tap is full of liquid before and after the liquid was run out. All experienced chemists operate the tap with their left hand, using the right hand to shake the vessel into which the liquid is being run. You should practise this until you are confident that you can control the tap to deliver one drop at a time. It is best to read the burette with a piece of white paper held behind it.

The burette is mostly used for titrations. It should be possible for you to titrate to the nearest drop. One drop is approximately 0.05 cm^3 so that if your successive titrations differ by more than 0.05 cm^3, there is something wrong with your technique of using the pipette or the burette, or in your way of judging the end point.

Readings

It is essential that every measurement is always written down straight away. Much the best method is to have a notebook which can be taken to and from the balance and kept by your side throughout the experiment. Sheets of paper inevitably get lost.

This brief introduction cannot hope to cover all the problems of technique which you will face because good technique is so much a matter of experience. If you run into difficulties seek the guidance of your teacher.

1: A Study of the Mineral Dolomite

The object of this series of experiments is to investigate the qualitative and quantitative composition of the mineral dolomite, or dolomitic limestone. Large mountain ranges in Italy and Austria (the Carinthian Alps) are made up of this material; it also occurs in the Cambrian limestones of the north of Scotland, and in the Permian beds from Tynemouth to Nottingham in England. It is also common in Canada and South Africa.

Dolomite has been used in the construction of some of our well-known public buildings, including the Houses of Parliament. It is also largely used as the raw material for the manufacture of the refractory linings of Bessemer converters.

Dolomites from various sources vary somewhat in quantitative composition, but all contain certain fundamental constituents. You will find out what they are and will then determine the percentages of the major ones in a sample of the mineral.

1.1 Investigation of the Metal Elements Present in a Sample of Dolomitic Limestone

Pure limestone is calcium carbonate. Some limestones contain magnesium carbonate as well as calcium carbonate and many limestones contain some iron(III). In this experiment you are to devise a scheme to find out whether a given sample of limestone contains magnesium and iron.

You are provided with solutions of calcium ions, magnesium ions, and iron(III) ions in dilute hydrochloric acid. Each is 0.1 molar with respect to the metal ions. You are also provided with three reagents A, B, and C. Firstly, you will build up your own scheme of analysis using these three reagents and then you will use the scheme you have developed to analyse your sample of limestone. You should work either on the semimicro scale or on the macro scale as recommended by your teacher.

Apparatus and Materials

Semimicro scale
Semimicro test-tubes in a test-tube rack
Beakers of distilled water for washing purposes

Teat pipettes
Glass rod to fit semimicro test-tubes
Micro Bunsen burner or beaker of boiling water to heat solutions in semimicro
  tubes
Centrifuge

Macro scale
A rack of test-tubes
Wash-bottle of distilled water
Glass rod
Filter funnel and filter papers
Bunsen burner

Solutions
Fe^{3+} 2.7 g of $FeCl_3,6H_2O$ in 20 cm³ bench 2 M hydrochloric acid and made up
  to 100 cm³
Ca^{2+} 1.0 g of $CaCO_3$ dissolved in 20 cm³ of bench hydrochloric acid and made up
  to 100 cm³
Mg^{2+} 0.84 g of magnesium carbonate, basic, dissolved in 20 cm³ of bench hydrochloric acid and made up to 100 cm³
A   100 cm³ of 5 M ammonium hydroxide containing 27 g of ammonium
  chloride
B   100 cm³ of A containing 16 g of ammonium carbonate
C   100 cm³ of B saturated with diammonium hydrogen phosphate

Litmus, or suitable pH papers
Powdered sample of dolomitic limestone

Procedure

Test each of the three metal ion solutions with the three reagents separately.
Note the formation of any precipitates and their colour and record the results
in the form of a table. There is no need to heat the solutions.

It is now necessary to use the data you have obtained to make up an analytical scheme. The problem is to detect one metal ion in the presence of the
others. Consider carefully the detection of magnesium in the presence of iron
and calcium. You have found that only one of the reagents gives a precipitate
with magnesium. (We will call this reagent X for convenience.) X also gives a
precipitate with iron and with calcium so that it will not be possible to detect
magnesium in a solution if iron and calcium are also in the solution. Is it
possible to remove iron and calcium from such a solution by precipitation
with one of the other two reagents? Try this out.

There are two points of technique which are important. Firstly, a precipitate
can be coagulated by warming it gently. It then filters or centrifuges much
more easily. Secondly, it is necessary to ensure that precipitation by a given
reagent is complete; do this by adding a little more of the reagent to the
filtrate.

The resulting solutions from iron and calcium can now be tested with reagent X. If they no longer give a precipitate with X the ions have effectively

been removed from the solutions and it will now be possible to use this technique for detecting magnesium in the presence of both iron and calcium.

Write out the scheme you have developed for finding out whether iron(III) and magnesium are present in solutions which are known to contain calcium. Make up a mixture of equal volumes of the solutions containing calcium, magnesium, and iron(III) ions and use it to find out if your scheme works. Finally dilute the solution of Fe^{3+} by factors of 50, 100, and 200 and find out the sensitivity of your test for iron. What is the minimum weight of iron per dm^3 which can be detected?

Now use your scheme to analyse the sample of limestone. Dissolve about 0.25 g in 20 cm^3 of approximately 2 M hydrochloric acid, warming gently. Remove any slight residue (likely to be silica) by centrifuging or by filtration. Use the clear solution as your sample for analysis and find out if it contains iron and magnesium.

Record of the Experiment

Give the results of all the tests carried out and state your conclusions about the composition of the limestone.

Write ionic equations for all the precipitation reactions which have occurred.

Your analytical scheme is entirely dependent on the different solubilities of various compounds. Which member of the following pairs do you think is less soluble in water?

> Calcium and iron(III) hydroxides
> Calcium and magnesium carbonates

1.2 Estimation of the Alkalinity of Dolomitic Limestone

Limestones may contain both carbonate ions and oxide or hydroxide ions. Both will react with acids and therefore form part of the alkalinity. Limestones will not dissolve in water, so that a solution cannot be directly estimated against an acid. Instead the limestone must be reacted with an excess of acid and the amount of excess acid determined by titration against a standard alkali.

Apparatus and Materials

250 cm^3 tall beaker	Balance accurate to 10 mg
Filter funnel	Approx. 2 M hydrochloric acid
Titration apparatus	0.10 M sodium hydroxide
250 cm^3 standard flask	Screened methyl orange
500 cm^3 standard flask	Powdered sample of limestone

Procedure

(a) Dissolve an accurately weighed quantity, about 1 g, of the limestone in a known volume, 25 cm³ by pipette, of roughly 2 M hydrochloric acid. You will need to warm to effect solution and you should avoid loss of acid spray by placing a funnel in the beaker. Do not boil, or warm for too long or some hydrogen chloride vapour will be lost. Filter and make up to 250 cm³ in a standard flask. You must take all precautions to ensure that all the solution from the beaker is transferred to the standard flask and this will involve a number of washings of all the apparatus involved. The accuracy of your result will depend very much on your technique here.

(b) Determine the molarity of the initial hydrochloric acid by diluting twenty times using volumetric apparatus and titrating the diluted solution against the 0.1 M sodium hydroxide; 10 cm³ portions are suitable for the titration and screened methyl orange should be used as the indicator.

(c) Determine the molarity of the excess hydrochloric acid in (a) by titration against the 0.1 M sodium hydroxide, using 10 cm³ portions and screened methyl orange as indicator as in (b).

Calculations

$$NaOH + HCl \longrightarrow NaCl + H_2O$$

$$\frac{\text{Volume of NaOH in titration}}{1000} \times \text{molarity of NaOH}$$

$$= \frac{\text{Volume of HCl in titration}}{1000} \times \text{molarity of HCl}$$

First calculate the molarity of the two solutions of hydrochloric acid which you titrated. Then calculate the number of moles of hydrogen chloride present in the initial 25 cm³ and in the 250 cm³ excess acid. Hence find the number of moles of hydrogen chloride used up by the limestone. Express the alkalinity as the number of moles of hydrogen chloride with which 100 g of the limestone will react.

Discussion

You will relate this result to the result you obtain in Experiment 1.3. This will enable you to say something about the composition of the limestone you are investigating.

1.3 Estimation of the Amount of Carbonate Ion in Dolomitic Limestone

When it reacts with an acid, the carbonate ion gives carbon dioxide. The volume of carbon dioxide will be related to the amount of carbonate ion which has reacted.

Apparatus and Materials

10 × 75 mm test-tube
25 × 150 mm test-tube fitted with rubber bung and right-angled glass tube
Glass gas syringe, 100 cm³

Balance accurate to 1 mg
Concentrated hydrochloric acid
Powdered sample of limestone
A marble chip

Procedure

Find the volume of carbon dioxide evolved from a given mass of powdered limestone. Weigh approximately 0.3 g of the powdered limestone in a small test-tube. Place approximately 15 cm³ of concentrated hydrochloric acid in a large test-tube. Saturate this acid with carbon dioxide by dropping in a small marble chip. It is best to break this chip up in a mortar but not powder it, so that it reacts at a reasonable rate. Why is it important to saturate the acid?

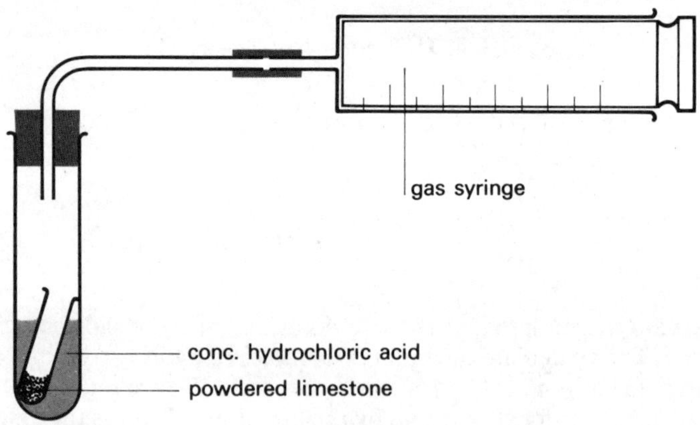

Figure 1.1

Place the small test-tube in the large one and connect to a gas syringe (see Figure 1.1). Do not allow the substances to react until you have tested for leaks. Mix the limestone with the acid and find the total volume of carbon dioxide evolved.

Record the laboratory pressure and temperature.

Calculation

Calculate the volume of 1 mole of gas under laboratory conditions. Hence find the number of moles of carbon dioxide evolved and the number of moles of carbonate ion which have reacted.

Express your result as moles of carbonate ion present in 100 g of the limestone.

Discussion

Compare this result with that of Experiment 1.2. This will enable you to say something about the composition of the limestone. Does it consist of a normal carbonate or a basic carbonate? Basic carbonates contain free oxide or hydroxide ions as well as carbonate ions.

2: Carbonate Titrations

2.1 A Study of Possible Indicators that can be used in Titrating Carbonates

In Experiment 1.2 you used screened methyl orange in the titration of a strong acid against a strong base. This experiment is concerned with the choice of a suitable indicator for the titration of a carbonate.

You will investigate four indicators: screened methyl orange, phenolphthalein, bromothymol blue, and thymol blue.

Apparatus and Materials

Plenty of test-tubes 10 × 75 mm   Screened methyl orange
2 teat pipettes   Phenolphthalein
0.10 M hydrochloric acid   Bromothymol blue
0.05 M sodium carbonate   Thymol blue
0.10 M sodium hydrogen carbonate   Titration apparatus

Procedure

First get a quick idea of the colour changes involved by testing the indicators in turn with 0.1 M hydrochloric acid, 0.05 M sodium carbonate, 0.1 M sodium hydrogen carbonate, and distilled water. Make a table of your results.

Now titrate 10 cm^3 portions of 0.05 M sodium carbonate against 0.1 M hydrochloric acid, using each indicator in turn. Shake very thoroughly during each titration. Record the full range of volume over which the indicator gave its colour change.

Record and Comment

Compare the titration results with the theoretical figures remembering that there are two possible reactions:

$$CO_3^{2-} + H^+(aq) \longrightarrow HCO_3^-$$

$$CO_3^{2-} + 2H^+(aq) \longrightarrow H_2O + CO_2$$

State clearly whether each of the indicators is satisfactory for the determination of both, only one, or neither of the two end points.

The pH ranges of the indicators are as follows:

Screened methyl orange	2.9–4.6
Phenolphthalein	8.3–10.0
Bromothymol blue	6.0–7.6
Thymol blue	1.2–2.8 and 8.0–9.6

From your results what do you think is the value of the pH at the end point of the two reactions?

2.2 The Potentiometric Titration of a Carbonate Solution

In a potentiometric titration the end point is judged by means of an instrument which will measure the electrical potential between two electrodes, one of which is a standard and the other is inserted in the solution. A pH meter is such an instrument and most commonly the electrode inserted in the solution is one which changes its potential with a change in the pH of the solution. A glass electrode is commonly used. In Experiment 2.1 you observed that when a carbonate is acted upon by an acid there are two distinct end points. These end points are at quite different pHs.

The purpose of this experiment is for you to observe the two end points by using a pH meter. The results will allow you to decide whether you would prefer to use a pH meter or an indicator in the estimation of a carbonate.

Apparatus and Materials

Titration apparatus and solutions as for Experiment 2.1
50 or 100 cm³ plastic beaker

pH meter
Solution of known pH to standardize the meter

Procedure

Set up a pH meter and standardize it against a solution of known pH, if necessary with the guidance of your teacher. Titrate a 10 cm³ portion of the 0.05 M sodium carbonate against 0.1 M hydrochloric acid and measure the pH after successive additions of 0.5 cm³. Between 4.0 and 6.0 cm³, and 9.0 and 11.0 cm³, measure the pH at intervals of 0.2 cm³. After 11.0 cm³ make only one further measurement at 15.0 cm³. Remember to mix thoroughly after each addition and take great care not to scratch the glass electrode. It is safer to use a plastic rather than a glass beaker as a titration vessel. It may be necessary to add distilled water to the 10 cm³ portion to cover the electrode.

Record and Comment

Choose a suitable method to present your experimental results. State the precise end points of the two reactions and explain which end point would be the most easy to judge using a pH meter.

From the results of Experiments 2.1 and 2.2 would you prefer to use an indicator or a pH meter in the estimation of a carbonate?

3: Polyprotic Acids

3.1 Analysis of Bath Salts

Some bath salts are simply washing soda, sodium carbonate decahydrate (Na_2CO_3, $10H_2O$). In concentrated solution they are quite caustic and harmful to the skin. Other bath salts are a compound of sodium carbonate and sodium hydrogen carbonate and are much milder to the skin. Most bath salts contain some perfume and possibly a little detergent. In order to analyse bath salts it is necessary to be able to determine a carbonate and a hydrogen carbonate separately.

In Chapter 2 you found that sodium carbonate gives two end points when titrated with an acid, corresponding to the reactions:

Stage 1 $$CO_3^{2-} + H^+(aq) \longrightarrow HCO_3^-$$

Stage 2 $$HCO_3^- + H^+(aq) \longrightarrow H_2O + CO_2$$

Overall $$CO_3^{2-} + 2H^+(aq) \longrightarrow H_2O + CO_2$$

You also found at what pH these two end points occur and what indicator is suitable for each end point.

The equation for the action of a hydrogen carbonate on an acid is:

$$HCO_3^- + H^+(aq) \longrightarrow H_2O + CO_2$$

Will the end point of this reaction correspond to the first or the second stage of the carbonate reaction?

When a mixture of sodium carbonate and sodium hydrogen carbonate is titrated using a pH meter or suitable indicators, the volume of hydrochloric acid required to complete the first stage can be used to calculate the amount of carbonate present. The difference between the volume of acid required for the second stage and that required for the first stage can be used to calculate the amount of hydrogen carbonate initially present. Thus both carbonate and hydrogen carbonate can be analysed together.

The problem is to find the molecular proportions of carbonate to hydrogen carbonate in the bath salts with which you are supplied.

Apparatus and Materials

Weighing bottle
Balance accurate to 10 mg
250 cm³ standard flask
Funnel
Titration apparatus

pH meter and standard pH solution for
  standardization
Screened methyl orange
Thymol blue
0.10 M hydrochloric acid
Bath salts

Procedure

Start by making a standard solution of the bath salts. A suitable amount will be about 2.5 g in 250 cm^3. Choose either the technique of Experiment 2.1 or 2.2 to determine the two end points of the reaction of this solution with 0.1 M hydrochloric acid.

Calculation and Comments

The volume of acid required to reach the first end point will allow you to find the molarity of the carbonate. The difference in volume of acid required for the second stage and that required for the first stage will allow you to calculate the molarity of the hydrogen carbonate.

If the ratio of these molarities is 1:1, the bath salts must be pure sodium sesquicarbonate (Na_2CO_3, $NaHCO_3$, $2H_2O$). Any ratio with the hydrogen carbonate proportion less than this must indicate that there is some free sodium carbonate in the bath salts. This means that they are not of very high quality.

Do you conclude that the bath salts you used were of high or poor quality?

3.2 Study of the Neutralization of Phosphoric Acid

Orthophosphoric acid, H_3PO_4, is a triprotic acid and will therefore form more than one salt with an alkali. This experiment is designed to study the stoichiometry of the different reactions of orthophosphoric acid with sodium hydroxide. A different indicator is used for the detection of each of the end points which correspond to the formation of the different salts of the acid. Write down the formulae of the salts you would expect to be formed.

Apparatus and Materials

Titration apparatus
0.10 M sodium hydroxide
0.05 M orthophosphoric acid

Thymolphthalein indicator
Bromocresol green indicator

Procedure

Titrate 25 cm^3 portions of 0.05 M orthophosphoric acid with 0.1 M sodium hydroxide, using bromocresol green as the indicator. From your results find the reacting mole ratio of sodium hydroxide with orthophosphoric acid using this indicator, and write a balanced equation for the reaction.

Titrate 25 cm^3 portions of 0.05 M orthophosphoric acid with 0.1 M sodium hydroxide, using thymolphthalein as indicator. Treat your results exactly as before, finally writing a balanced equation for the reaction.

Consult a reference book to find out the pH ranges of both the indicators you have used.

4: Acetylsalicylic Acid

4.1 Analysis of Aspirin Tablets— A Consumer Survey

In all pharmaceutical preparations, the manufacturer is required by law to state on the packaging the maximum amount of each active ingredient present. In many preparations this will form only a small percentage of the pill or tablet as a whole. In many cases, the tablet would simply disintegrate into a powder unless additives were put in to assist in making the ingredients cohere into tablet form.

In this experiment you will carry out a consumer survey on the aspirin content of a number of commercial preparations, to see whether the manufacturer's claims are justified.

Acetylsalicylic acid (aspirin) can be readily hydrolysed by sodium hydroxide into the sodium salts of two weak acids, ethanoic (acetic) acid and salicylic acid. In this experiment, the hydrolysis is effected using an excess of sodium hydroxide, the excess being later found by titration with standard sulphuric acid. The equation for the hydrolysis reaction is:

$$CH_3.COOC_6H_4.COOH + 2NaOH \longrightarrow$$

$$CH_3.COONa + HO.C_6H_4.COONa + H_2O$$

Apparatus and Materials

Titration apparatus

Various commercial preparations containing acetylsalicylic acid (e.g. aspirin, Aspro, Disprin)

Phenol red indicator

1.0 M sodium hydroxide

0.05 M sulphuric acid

Procedure

Weigh accurately a definite number of aspirin tablets (two or three tablets, not more than 1.5 g) into a 250 cm³ conical flask. Initiate the hydrolysis of the aspirin by adding 25 cm³ of 1 M sodium hydroxide by pipette, diluting with approximately the same volume of distilled water. Warm the flask over a tripod and gauze for ten minutes to complete the hydrolysis.

Cool the reaction mixture and transfer with washings to a 250 cm³ standard flask. Dilute to the mark with distilled water and then ensure that the contents of the flask are well mixed by repeated shakings.

Titrate 25 cm^3 portions of the diluted reaction mixture with 0.05 M sulphuric acid, using phenol red as indicator. This indicator (with pH range 6.8–8.4) is most suitable for this titration because of the presence of the salts of two weak acids. Explain why.

Calculate the weight of acetylsalicylic acid in each tablet and compare your result with the manufacturer's specification. Compare your result with those obtained by students analysing different makes of aspirin tablets.

5: Determination of Calcium

5.1 Determination of Calcium Gravimetrically as its Oxalate

In Chapter 1 you determined the percentage of carbonate in dolomite and you have also found that it contains magnesium as well as calcium. The next stage in the analysis is to determine the percentage of calcium. You will use a technique which, in contrast to volumetric analysis, involves only weighing. This is known as gravimetric analysis and it can be extremely accurate.

It is necessary to convert all the calcium into a compound which can be completely separated from the rest. This involves choosing a suitable insoluble salt of calcium of which the magnesium salt is soluble. The oxalate, CaC_2O_4, H_2O, is just such a compound. The precipitated calcium oxalate must be in a suitable state for complete and rapid separation by filtration and in the first part of the experiment you will investigate these conditions.

Apparatus and Materials

Test-tubes and rack
Filter funnel
Gooch crucible and asbestos, or sintered glass crucible
Filtration flask for above
Oven
Desiccator
Balance accurate to 1 mg
600 cm³ beaker
Evaporating basin to cover the beaker

Glass rod with rubber tip
Approx. 0.5 M calcium chloride solution
Ammonium oxalate solution, about 40 g per dm³
Approx. 5 M hydrochloric acid
Approx. 5 M ammonia
Methyl red indicator
Powdered sample of limestone

Procedure

(a) Mix equal volumes of calcium chloride solution and ammonium oxalate solution in a test-tube. Filter half immediately, observing the nature of the filtrate. Boil the other half, allow to stand for half an hour and then filter. Compare this filtrate with that of the first half. From the results of these experiments, decide which is the better technique to adopt.

(b) Prepare a Gooch crucible with asbestos wool. You will need your teacher's instructions for the technique required. A sintered glass crucible can be used instead of a Gooch crucible. Dry the crucible in an oven at 105 °C and allow to cool in a desiccator before weighing. Weigh out accurately about 1 g of the powdered limestone and transfer it to a 600 cm³ beaker. Cover the beaker with an evaporating basin or watch-glass to avoid loss by spray and run in approximately 20 cm^3 of 5 M hydrochloric acid. Boil to dissolve and to remove carbon dioxide and then add approximately 200 cm^3 of distilled water. Bring to the boil again, remove the source of heat and add 50 cm^3 of the ammonium oxalate solution. Add sufficient methyl red to colour the solution faintly red and then add 5 M ammonia solution a little at a time until the solution is just neutral. This is judged by the methyl red changing from red to orange. There must not be excess ammonia or iron(III) hydroxide may be precipitated if any iron(III) ion is present in the limestone. This means that at the furthest the methyl red should just have become yellow with the last drop of ammonia added. Allow to stand for half an hour and then transfer all the precipitate to the Gooch crucible. The technique for this transfer is extremely important and you should seek help from your teacher. Wash the precipitate in the crucible with distilled water and dry it for half an hour in an oven at 105 °C. Allow to cool in a desiccator before finally weighing.

Calculation

The formula of the product is CaC_2O_4, H_2O. From the weight of the product determine the weight of the calcium in the original limestone weighed out. Express your result as the number of moles of calcium per 100 g of limestone.

5.2 Determination of Calcium by a Combined Gravimetric and Volumetric Method

In this experiment you will carry out the same analysis as in Experiment 4.1 but you will use a different technique. The calcium oxalate is precipitated as before but it is then dissolved in acid. The oxalate ion is a strong reducing agent and can be determined by titration against standard potassium permanganate.

$$16H^+ + 5C_2O_4^{2-}(aq) + 2MnO_4^-(aq) \longrightarrow 10CO_2(g) + 2Mn^{2+}(aq) + 8H_2O$$

The reaction is rather slow at room temperature but rapid above 60 °C.

Apparatus and Materials

Balance accurate to 1 mg
600 cm^3 beaker
Evaporating basin to cover the beaker
Glass rod with rubber tip
Titration apparatus
250 cm^3 standard flask
Ammonium oxalate solution, about
   40 g per dm^3

Approx. 5 M hydrochloric acid
Approx. 5 M ammonia
Methyl red indicator
0.02 M potassium permanganate
Approx. 1 M sulphuric acid
Powdered limestone

Procedure

Obtain the precipitate of calcium oxalate just as in Experiment 4.1 and allow it to stand for half an hour. Filter it in a normal filtration apparatus, *making sure that all the precipitate is transferred to the filter paper*. Wash it thoroughly on the filter paper with distilled water and choose a suitable test to find out whether the washings still contain oxalate ion. If they do, wash a little more. Make a hole in the filter paper with a glass rod and wash all the precipitate into a 250 cm^3 measuring flask with 1 M sulphuric acid. Make up to the mark with the 1 M sulphuric acid. Titrate 10 cm^3 portions against 0.02 M potassium permanganate, heating to approximately 70 °C before each titration. The end point is indicated by the first sign of a permanent pink colour arising from excess permanganate.

Calculation

$$5 \text{ moles of } C_2O_4^{2-} \text{ react with 2 moles of } MnO_4^-.$$

Hence

$$\frac{\text{volume of oxalate} \times \text{molarity of oxalate}}{\text{volume of permanganate} \times \text{molarity of permanganate}} = \frac{5}{2}$$

Determine the molarity of the oxalate solution and hence the number of moles of calcium oxalate precipitated. From this obtain the number of moles of calcium in the limestone originally weighed out. Express your final result as moles of calcium in 100 g of the limestone.

Compare your result in this experiment with that in Experiment 4.1. State, with reasons, which result you believe to be the more accurate.

6: Hardness in Natural Water

6.1 Determination of Calcium and Magnesium in Natural Water

Limestones and magnesium limestones dissolve to some extent in rain water so that spring and river water and, of course, sea water, contain calcium and magnesium ions. Sea water is an industrial source of magnesium. Hardness in natural waters is caused by calcium and magnesium ions. It is therefore of considerable importance to be able to determine the concentrations of these ions in solution.

The object of this experiment is to find the concentration of magnesium and calcium ions in a sample of natural water. You may be working with sea water or with spring water. The concentration of the ions in sea water is much greater than in spring water, and so the sea water has to be diluted before carrying out the determination.

For the titration you will be using ethylenediamine-tetraacetic acid, commonly abbreviated to EDTA. This reagent forms complexes with a number of metal ions. Its formula is best represented as the zwitter ion

$$^{-}O_2C\text{---}CH_2 \qquad\qquad\qquad CH_2\text{---}CO_2^{-}$$
$$H\text{---}N^{+}\text{---}CH_2\text{---}CH_2\text{---}N^{+}\text{---}H$$
$$^{-}O_2C\text{---}CH_2 \qquad\qquad\qquad CH_2\text{---}CO_2^{-}$$

A M^{2+} ion forms six ligands—four with the four CO_2^{-} groups and two with the two N atoms, so that one metal ion complexes with one EDTA molecule:

$$H_2Y^{2-} + M^{2+} \rightleftharpoons MY^{2-} + 2H^{+}$$

Complexes formed with M^{2+} ions are very stable, but, even so, there must be some metal ions in equilibrium with the complex. If, then, an anion, such as OH^{-}, which forms a very insoluble compound with the free metal ion is present in the solution, it will disturb the equilibrium and the complex will be broken down, with precipitation of the insoluble compound. Thus, at a sufficiently high concentration of OH^{-} the complex of Mg^{2+} with EDTA would be decomposed and $Mg(OH)_2$ would be precipitated.

You have already found (Experiment 1.1) that $Ca(OH)_2$ is much more soluble than $Mg(OH)_2$, so that at the same concentration of OH^{-} the Ca^{2+} complex with EDTA would not be broken down.

The indicators used are also complexing agents of complicated structure, which, fortunately, you need not know. The principle is that the complexes they form with metal ions are less stable than those formed with EDTA. Thus, if the indicator, which we may call *In*, is added to a solution of Mg^{2+} the complex $Mg^{2+}In$ is formed, and it has a certain colour. As EDTA is added the Mg^{2+} is removed from the $Mg^{2+}In$ to complex with the EDTA:

$$Mg^{2+}In + EDTA \rightleftharpoons Mg^{2+}(EDTA) + In$$

The equilibrium lies far to the right so that the colour of the liquid is that of the non-complexed indicator.

These indicators are not very stable in solution and are therefore normally kept as solids. They are mixed with an inert salt so that they can be added in small enough quantities for each titration. Those you will be using here are called Patton and Reader's indicator and Solochrome black (sometimes called Eriochrome Black T).

Apparatus and Materials

Titration apparatus
250 cm³ standard flask
Filter funnel
2 teat pipettes
Sea water
Spring water
0.01 M EDTA solution

8 M potassium hydroxide (this solution has pH greater than 12.5)
Ammonia/ammonium chloride buffer of pH 10
Patton and Reader's indicator
Solochrome black indicator

Procedure

Dilute sea water by a factor of five using normal volumetric apparatus. Take 50 cm³ of spring water by pipette or 10 cm³ of diluted sea water. Find out how much in a teat pipette is approximately 2 cm³. Take the solution of pH > 12.5 and add 4 cm³ of it to the 50 cm³ portion or 2 cm³ to the 10 cm³ portion. Take care when using this reagent as it is highly caustic and must not be spilt on your skin or on the bench. If by any chance you do get some on your fingers, wash it off with water and put on some dilute ethanoic (acetic) acid and wash again. Use a spatula to add approximately 0.01 g of solid Patton and Reader's indicator. You may need to weigh this the first time but you can judge it afterwards. Titrate with the EDTA solution. It is necessary to shake thoroughly throughout the titration. The end point is when the purple colour has fully changed to a pure blue. It is not a very easy end point to judge and you may find it helpful to use your first titration with a few drops of excess EDTA as a colour match for later titrations. At this pH all the magnesium is precipitated as magnesium hydroxide and does not react with the EDTA. Hence this titration estimates the calcium alone.

Repeat the same procedure in order to estimate the calcium and magnesium together. This time use the buffer solution of pH 10 instead of KOH of pH > 12.5 and use Solochrome black as indicator. It is important to have sufficient of this indicator and approximately 0.08 g should be used.

Calculation

One mole of calcium ion or of magnesium ion reacts with one mole of EDTA. Hence the molarity of the sea water or spring water can easily be found.

Using your result for sea water, calculate the total volume of sea water which will be required to yield a kilogram of magnesium.

Hardness of water is sometimes expressed in degrees. A degree of hardness is a concentration of calcium and magnesium equivalent to 1 g of calcium carbonate in 100 000 g of water. As the molecular mass of calcium carbonate is 100, the degree of hardness is equivalent to a total calcium and magnesium molarity of 10^{-4}. Anything in excess of 20 degrees of hardness would be considered a really hard water. Is your sample of spring water very hard?

6.2 The Solubility of Calcium Sulphate at Different Temperatures

You are to find out whether the solubility of calcium sulphate increases or decreases with an increase of temperature. You should do this by determining the solubility at room temperature and at a temperature very near to the boiling point of the solution.

Apparatus and Materials

Titration apparatus
2 × 250 cm³ beakers
Filter funnel
Water saturated with calcium sulphate

0.01 M EDTA solution
Buffer solution, pH greater than 12.5
Patton and Reader's indicator

Procedure

Design your own experiment to carry out this investigation based on your experience of the estimation of calcium in solution by titration with EDTA, which you used in Experiment 6.1. To ensure that a solution is saturated at any stage there should be excess of the solid calcium sulphate present. In order to choose a suitable volume of solution for titration you should know that the concentration of a saturated solution of calcium sulphate is of the order of 0.01 M.

When a buffer of pH > 12.5 is added to moderately concentrated solutions of calcium ions there may be a slight precipitate of calcium hydroxide. This has to dissolve during the titration and this involves very thorough shaking. After the first titration this problem is avoided by running in the EDTA to within 1 cm³ of the end point before the buffer and indicator are added. The titration is then completed in the normal way.

Report

Make a report on your investigation and state whether the solubility of calcium sulphate increases or decreases with rise in temperature. What can you say about ΔH for the reaction?

$$CaSO_4(s) + H_2O \rightleftharpoons Ca^{2+}(aq) + SO_4^{2-}(aq)$$

6.3 The Solubility of Limestone in Waters Saturated with Carbon Dioxide

In Experiment 6.1 you discovered that natural water often contained calcium and magnesium ions. It is very often the limestones which are a source of these ions but neither calcium nor magnesium carbonate are appreciably soluble in water. It is said that the solubility is partly the result of the carbon dioxide present in rain water. In this experiment you are to investigate whether this seems to be a reasonable idea.

Apparatus and Materials

As for Experiment 6.2 with the addition of water saturated with carbon dioxide and powdered calcium carbonate

Procedure

You should do this by making a saturated solution of calcium carbonate in water which has been saturated with carbon dioxide. Compare the concentration of calcium ions in it with that in a saturated solution of calcium carbonate in pure water.

The estimation of the concentration of calcium ions should be based on the work you have done in Experiments 6.1 and 6.2. You should plan the experiment entirely yourself. You will need to find a reasonable volume for titration purposes by experiment.

Report

Write a full report on your investigations.

6.4 The Effectiveness of Water Softeners

In Experiments 6.2 and 6.3 you found that natural water may contain calcium and magnesium ions. The resulting water is said to be hard. Hard waters react with soaps to give a scum. Various softeners are sold which can be added to the water before it is used. In this experiment you are going to find out which of two softeners is the best to buy. The method you will use will depend on a titration but the means of judging the end point does not involve the use of an indicator. It is in fact a good illustration of the basic requirement of any titrimetric analysis, i.e. that there must be some means of determining the point at which the reaction is completed. Any definite change in property at this point will do. In this case the change in property is the ability to form a lather. As long as there are any calcium and magnesium ions in the solution soap reacts with them to form a precipitate and there is no soap to form a lather. The moment a lather can be formed all the ions causing hardness have been precipitated.

Apparatus and Materials

25 or 50 cm³ graduated cylinder
Titration apparatus
2 × 250 cm³ conical flasks fitted with corks
Balance
Standard soap solution (10 g of soft soap should be dissolved in 250 cm³ of distilled water and 250 cm³ of colourless methylated spirit added)

Calgon water softener
Borax water softener
Sample of hard water (if this is not available locally it can be made by diluting 50 cm³ of limewater to 500 cm³ and adding about 0.13 g of hydrated magnesium sulphate)

Procedure

Measure into a conical flask 25 cm³ of distilled water by means of a graduated cylinder. Add soap solution from the burette in 0.5 cm³ portions. After each addition, cork the flask and shake thoroughly for at least ten seconds. Judge whether a good lather has formed by allowing the flask to stand for a full minute and seeing if a lather remains. For a good lather it should at least cover the whole surface at this stage. Find out how much soap solution is required to give a good lather. Distilled water may be considered to have no hardness, so that you have a standard to use in judging whether a given sample of water has been completely softened. Carry out the same estimation with your sample of hard water.

Now it is necessary to find out how much of each softener is required to completely soften the water. This is conveniently done by first making solutions of the two softeners used. Dissolve 1 g of Calgon in 100 cm³ of distilled water and 5 g of Borax in the same quantity of water. Add each in turn in 1 cm³ quantities to 25 cm³ of the hard water. Each time find out how much soap solution is required to give a good lather. Hence find the minimum quantity of the two softeners which will completely soften a 25 cm³ portion of the water.

Record and Report

Estimate how much it will cost with the two softeners to completely soften 10 kg of the hard water and write a short report on each of them.

Discussion

Why is it satisfactory to measure out the water with a graduated cylinder, instead of using a pipette?

Explain whether you regard this as a very accurate way of finding the hardness of a water sample.

7: Ion Exchange

7.1 Determination of the Total Cation Concentration of an Aqueous Solution by Ion Exchange

Most natural waters, including sea water, contain calcium, magnesium, sodium, and potassium ions in considerable quantities but only traces of other ions. In Experiment 6.1 you have discovered how to determine the concentrations of the Ca^{2+} and Mg^{2+} by titration with EDTA but the direct determination of Na^+ and K^+ is extremely difficult, except by the use of a flame photometer. However, if the total cation concentration can be determined and the concentrations of Ca^{2+} and Mg^{2+} are known, the total Na^+ and K^+ concentrations can be obtained by difference.

In this experiment you will first investigate ion exchange and then make use of the knowledge you have gained to determine the total cation concentration of a sample of natural water. You will use an ion exchange resin in what is known as its acid form.

Apparatus and Materials

100 cm³ conical flasks
Teat pipettes
Test-tubes
Glass column approximately 1.5 cm × 25 cm fitted with a tap at the bottom
Tap funnel to deliver solutions to the column
5 calibrated beakers, or 5 ordinary beakers and a 25 cm³ measuring cylinder
500 cm³ standard flask
Titration apparatus

30 cm³ of resin, Zeo Karb 225 in its acid form
0.01 M sodium hydroxide
0.01 M hydrochloric acid
Approximately 0.5 M sodium chloride, potassium chloride, and magnesium chloride
Calcium hydroxide solution
Methyl red indicator
Universal indicator
Litmus paper

Procedure

To 25 cm³ of distilled water in a *very clean* flask add about 10 drops of 0.5 M sodium chloride. Withdraw a small sample and determine its pH with universal indicator. Now add a little ion exchange resin to the solution in the flask and shake thoroughly. Test a sample with the universal indicator. Repeat the experiment with potassium chloride, calcium chloride, and magnesium chloride, in each case finding out the effect of the addition of the resin on the pH. Also find out the effect of the resin on the pH of distilled water. (The resin you use will probably be kept under distilled water.) About one-quarter of a spatula-full will be sufficient for each experiment. Draw conclusions about the exchange of ions which must have taken place between the solutions and the resin.

Take a further 25 cm³ of distilled water and add 10 drops of calcium hydroxide solution to it. Test its pH before and after the addition of a little resin. Does it make any difference if more resin is added? Account for the result of this experiment as completely as you can, consulting a textbook if necessary.

You will now have realized that this ion exchange resin will quantitatively exchange all cations for their equivalent in hydrogen ions. Determination of the concentration of hydrogen ions after a solution has been treated with the resin will allow the total concentration of cations in the solution to be measured.

Prepare an ion exchange column with about 25 cm³ of resin following the instructions from your teacher or from a suitable textbook. The apparatus is shown in Figure 7.1. Check that the eluent contains no free hydrogen ions when distilled water is run through the column. If you are using sea water dilute it by a factor of fifty, using normal volumetric apparatus. If you are

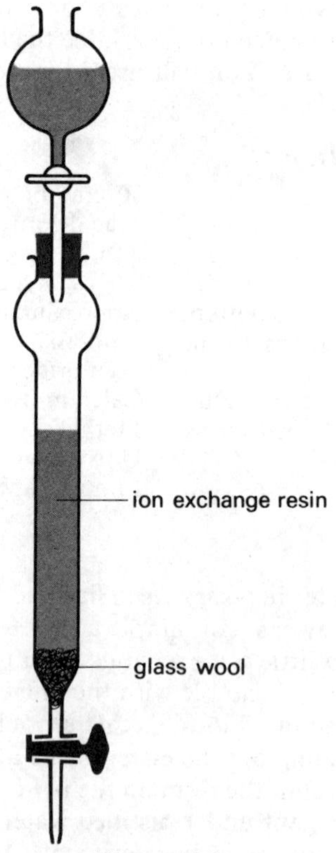

Figure 7.1

using fresh water there will be no need to dilute it. Run the water you are testing through the column at a rate of about 10 cm^3 per minute. Collect the first 50 cm^3 of the eluent and from then onwards 25 cm^3 fractions. Do not let the column run dry. Titrate a 10 cm^3 portion of each fraction against 0.01 M sodium hydroxide, using methyl red as indicator. You are titrating a very dilute solution and you should work out a good technique for obtaining a really good end point. The pH range of methyl red is 4.1–6.0. Continue to collect fractions until two successive titres agree. Why should the first fractions give a decidedly lower figure?

Your original sample of water before it has been through the column may be alkaline. Why is it necessary to determine its alkalinity? Do this by titrating it against 0.01 M hydrochloric acid, using methyl red as indicator.

Hence determine the number of moles of hydrogen ion produced when a litre of the sample of water is fully treated with the resin.

Calculation

Remember that cations with a double positive charge such as Ca^{2+} and Mg^{2+} will give two moles of hydrogen ion for every mole of cation which is passed through the column. Assuming that your sample of water contains only calcium, magnesium, sodium, and potassium as cations and using your result from this experiment and that from Experiment 6.1, determine the total potassium and sodium ion concentration of the natural water, expressing your result in moles per dm^3.

Recovery of Resin

Your teacher may wish you to convert back all the resin you have used into its original form. If so, do not discard it but follow instructions.

8: Analysis of Brass

8.1 Detecting the Elements Present in Brass

In the next four experiments you will investigate the composition of a brass. Brass is basically an alloy of copper and zinc, but other metals, including lead, tin, and nickel, may be present. We shall limit our analysis to the three most common constituents, copper, zinc, and lead. A small amount of brass is taken into solution in concentrated nitric acid. Lead can be detected by means of its chloride, which is only slightly soluble. Copper and zinc are detected chromatographically.

Apparatus and Materials

Tanks and paper for chromatography
Sample of brass
Capillary tubes
Spray for locating reagents
0.01 M solution of copper(II) ions in 2 M hydrochloric acid
0.01 M solution of zinc ions in 2 M hydrochloric acid

Solvent 1 (acetone–conc. hydrochloric acid–water (86–6–8))
Or
Solvent 2 (diethyl ether–methanol–water – concentrated hydrochloric acid (50–30–15–5))
Rubeanic acid (0.1 g in 100 cm^3 ethanol)
8-Hydroxyquinoline ('oxine') solution (0.5 g in 100 cm^3 of an 80:20 mixture of ethanol and water)

Procedure

(a) Place a small piece of brass in a beaker in a fume chamber and cover it with concentrated nitric acid. Add just sufficient nitric acid to dissolve all the brass.

(b) Carry out a simple test for the presence of lead(II) ions by adding a few drops of concentrated hydrochloric acid to a little of the solution from (a). Explain what happens in the test.

(c) The colour of the solution would suggest the presence of copper. However, paper partition chromatography will be used to separate the other metal ions present in the solution. If you are not already acquainted with it, study the theory of this most important analytical technique by consulting a suitable text. Put simply, separation of different components in a mixture occurs because the components have different solubilities in two solvents. These are the water held on the paper (stationary phase) and the solvent moving through the paper (moving phase). The repeated partition between these two solvents which occurs many times through the length of the paper leads to a successively greater separation of the components of the mixture. Clearly, if the components have similar solubilities in both solvents, little or no separation will occur.

The technique may be extended to substances which are colourless in solution, if after passing the solvent through the paper, it is sprayed with a suitable reagent which will react with a component of the mixture to give a coloured compound, which will then appear on the paper as a spot.

You are supplied with a tank and chromatography paper cut to fit it. Pour sufficient of the solvent supplied into the tank so as to cover the bottom. Replace the lid while preparing the paper, so that the atmosphere in the tank will become saturated with the solvent vapour. The preparation of the paper should be done most carefully to avoid contamination, using the following technique. Place a large sheet of paper on the bench top, and lay the chromatography paper on top of this. Wherever possible touch only the edges of the paper. Mark a faint *pencil* line (do not use a ball-point pen) across the bottom of the paper, between 2 and 3 cm from the edge. Mark four origins along this line at equally spaced intervals, avoiding marking too near to the edges of the paper. Mark each origin with a suitable code, in pencil. Apply each sample by putting one drop of the solution from a capillary tube on to the pencil-marked origin. Take care not to put too large a sample on the paper, and only use each capillary tube once to avoid contamination. The arrangement of the spots in this experiment is indicated in Figure 8.1. The spots should be dried with a hot air dryer, or allowed to dry out at room temperature.

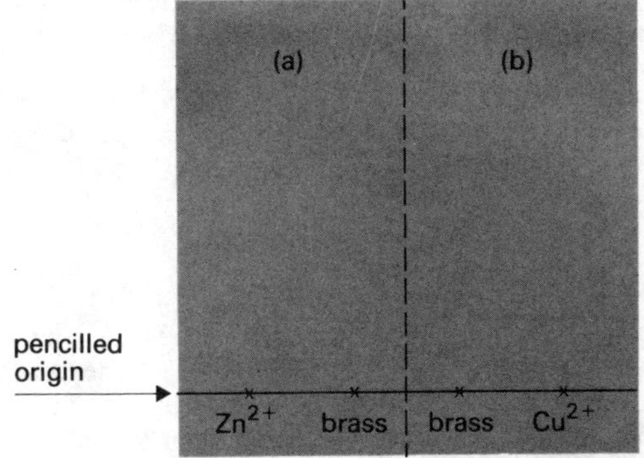

Figure 8.1

Now bring the two edges of the paper together to form a cylinder with the starting line at one end. Fasten the edges together with plastic clips or with paper clips, or with a paper stapler. If the edges of the paper overlap, considerable unevenness results in the flow of the solvent through the paper.

Lower the paper cylinder, spotted end first, into the tank and stand it carefully on the bottom. Make sure that it is not touching the sides of the tank. Replace the lid on the tank.

After the solvent front has travelled about half of the length of the paper, remove the chromatogram from the tank, mark the position of the solvent front with a pencil, open it out and dry it with the hot air dryer. Cut the paper in half vertically between the two origins of the brass samples (i.e. along the dotted line in Figure 8.1). The samples should now be sprayed with appropriate reagents which will indicate the presence of one or other of the metal ions.

Spray lightly the part of the paper marked (a) with oxine reagent, followed by exposure to vapour of 0.88 ammonia. Then view the paper under a u.v. lamp. Zinc gives a yellow fluorescence near the solvent front. Copper may also be shown up. Spray lightly the part of the paper marked (b) with the rubeanic acid reagent, followed by exposure to 0.88 ammonia. Copper and nickel may be detected by this reagent.

Identification of the different components of the brass is made by inspecting the relative positions of the spots on the paper. Identification is usually completed by measuring the R_f value, that is, the ratio of the distance travelled from the origin by the substance, compared to the distance travelled by the solvent front. Measure the R_f values for all the substances detected on your chromatogram.

8.2 Estimation of Copper and Zinc in Brass by a Volumetric Method

This experiment consists essentially of three parts. First, a weighed amount of the brass is dissolved in nitric acid; second, the copper content of the sample is found; and third, the zinc content is found.

Apparatus and Materials

Titration apparatus	0.05 M EDTA solution
0.05 M sodium thiosulphate	Buffer pH 6 (hexamethylenetetramine)
0.5 M potassium iodide	Xylenol orange indicator
M potassium thiocyanate	Sample of brass
Starch indicator	

Procedure

(a) Weigh accurately about 3 g of brass. Dissolve it in the minimum quantity of concentrated nitric acid in a 250 cm³ beaker. Transfer the solution, with washings, to a 250 cm³ standard flask. Make up to the mark with distilled water and shake the flask well.

(b) The determination of copper depends on the fact that copper(II) ions react with iodide ions in the following way:

$$2Cu^{2+} + 4I^- \longrightarrow 2CuI + I_2$$

The free iodine liberated may then be estimated by the standard methods of iodimetry, using sodium thiosulphate (see Experiment 10.1 for details). This reacts with iodine to form sodium tetrathionate and sodium iodide.

$$I_2 + 2S_2O_3^{2-} \longrightarrow S_4O_6^{2-} + 2I^-$$

To 25 cm³ of the brass solution in a conical flask, add sodium carbonate solution until a permanent precipitate is just obtained. Now dissolve the precipitate in the least possible quantity of dilute ethanoic (acetic) acid. Add 10 cm³ of 0.5 M potassium iodide solution, shaking well. Copper(I) iodide is at once precipitated, and free iodine is liberated. Titrate the liberated iodine against the standard 0.05 M sodium thiosulphate. When the iodine colour begins to fade, add starch indicator. When the starch colour fades, add about 10 cm³ of M potassium thiocyanate which will revive the starch colour. Then titrate *rapidly* to the end point. Repeat to obtain two consistent titres.

(c) Zinc is determined by titration with EDTA, which is used as a standard solution of the disodium salt, represented as Na_2H_2Y. It forms a complex with the zinc ions present in the solution:

$$Zn^{2+} + H_2Y^{2-} \longrightarrow ZnY^{2-} + 2H^+$$

Clearly the success of the method depends on the prior removal of all copper(II) ions as copper(I) iodide before the zinc estimation is begun. Hence you will use the titration mixture left from the determination of copper.

Add to a titration residue from (b) approximately 4 g of solid hexamethylenetetramine. This buffers the solution at pH 6. Then titrate with 0.05 M EDTA, using a pinch of solid xylenol orange as indicator. The colour change is from red to yellow at the end point. Repeat to obtain two consistent titres.

Calculations

Use the equations given above to calculate the molarities of copper and zinc in the brass solution. Calculate the percentage composition of the brass. Compare your results with the actual mass of brass originally taken.

Questions

1. How may the differences between your results and the original mass of the brass sample be accounted for?
2. Why does the addition of I^- to Cu^{2+} give a precipitate of Cu(I)I?

3. Why is it necessary to buffer the solution in (c)? Carry out some test-tube experiments to find out what happens if the solution is not buffered.
4. Why is it necessary to remove all Cu^{2+} ions before determining Zn^{2+}?
5. What is hexamethylenetetramine and how does it act as a buffer?

8.3 Estimation of Copper in Brass by Colorimetry

Method 1

In this experiment, you will find the concentration of a coloured solution by the method of colour matching. A basic assumption of the method is that two solutions of the same substance having identical colour intensities will be of the same concentration. In this very simple method, two identical tubes are used, and small amounts of the standard solution removed until the colours of the two solutions are identical when viewed through the whole height of the column of solution. Then the heights of the solutions in the two tubes will be in the ratio of their concentrations.

Apparatus and Materials

2 Nessler tubes (or matched test-tubes) Sample of brass
White light source 0.5 M copper(II) nitrate solution

Procedure

Introduce the standard 0.5 M copper nitrate into one tube, up to the mark on the glass. If any of the solution of brass in nitric acid remains from Experiment 8.2 use it in the other tube, otherwise make up a fresh solution (weigh accurately about 3 g of brass, dissolve in a minimum of concentrated nitric acid, and make up to 250 cm³).

Wrap white paper round each Nessler tube so that the sides of the tube are completely covered, just leaving the top and bottom open. Fasten the paper with sellotape.

View the two tubes together, looking down on to a white light source. Using a clean teat pipette, remove small quantities of the standard solution into a beaker, until the colours in the two tubes appear to be of identical intensity. If too much of the standard is removed at once, simply add drops from the beaker back into the tube until an exact match is obtained. Measure the heights of solution in the respective tubes. From this ratio and the known molarity of the standard solution, calculate the molarity of the brass solution. Use this result to calculate the mass of copper in the sample of brass taken, and thus the percentage of copper in the brass.

Method 2

In this more sophisticated method, the rather poor ability of the human eye to distinguish between differences in colour intensity is replaced by a sensitive photoelectric cell, linked in turn to a meter. In this way, with suitable scale calibration, a direct value may be obtained for the intensity of colour of the solution. A suitable filter must be chosen from those provided, specifically to deal with the copper solutions being tested.

Apparatus and Materials

Photoelectric colorimeter (or other comparator)

Matched glass tubes
0.5 M copper(II) nitrate solution

Procedure

Initially, obtain a calibration graph by taking a number of readings on the instrument using standard solutions. This is best achieved by taking a first reading with 0.5 M copper nitrate, and then with successively diluted solutions down to about 0.01 M. Take eight or ten readings for different dilutions and plot them on graph paper.

Then introduce the copper nitrate solution from the brass sample into the colorimeter in a matched tube and take a meter reading. Interpolation from the calibration graph will give the molarity of the copper solution. Now work out the percentage of copper in the original brass sample. Compare the result with that obtained by Method 1 and with that of Experiment 8.2.

Questions

1. What law of light absorption is used in making the statement on page 34, 'The heights of the solutions in the two tubes will be in the ratio of their concentrations'?
2. What special difficulties are encountered with the colorimetric method? Consider, for example, whether you could use this method if the brass contained nickel.
3. State, giving your reasons, what you consider to be the best of all the methods you have used for the determination of copper in brass.

9: Gas Analysis

9.1 Investigation of the Products of Burning of a Candle using Gas Analysis

It is frequently necessary to be able to determine the percentage of various gases in a mixture. In many instances this can be done by the measurement of some physical property. How many such methods can you think of? Often chemical absorption of the constituent gases one by one is the method employed and this is the type of analysis you will use in this experiment.

Apparatus and Materials

Bell jar with rubber bung
Glass trough or large bowl
Evaporating basin
Gas syringe, glass or plastic
2 disposable syringes with needles
20 cm³ gas burette with tap and suba-seal cap
Rubber tubing to fit gas syringe
Eye shields

Candle
Potassium hydroxide solution, saturated and kept under liquid paraffin
Alkaline pyrogallol kept under liquid paraffin
Plastic bowl containing approximately 2 per cent sulphuric acid (a safety precaution)

Procedure

Float a candle in an evaporating basin in a large trough or plastic bowl which contains water. Light the candle and lower a bell jar over it with the bung open. As soon as the water levels inside and outside the jar are equal, put the bung in firmly. Fully observe what happens.

You will probably recall this experiment from your early chemistry. It is often used as a means of showing that only about one-fifth of the air will support combustion. Does the result of the experiment you have just carried out indicate this?

It must be immediately apparent to you that the rise in the level of water is quite insufficient to account for one-fifth of the air being used up. What is the reason for this? Is it that all the oxygen has not been used? Is it that there is a lot of carbon dioxide present in the product which occupies the space of the oxygen used? Why did the candle go out anyway? You will attempt to answer these questions by analysing the gas in the jar after the candle has gone out.

Carbon dioxide will dissolve in alkalis. Oxygen, nitrogen, and the noble gases do not. Pyrogallol, 1,2,3-trihydroxybenzene, when alkaline, is an extremely strong reducing agent and will react with all the oxygen present in a gas mixture. If a measured volume of gas is shaken first with potassium hydroxide and then with alkaline pyrogallol, the percentage of carbon dioxide and oxygen can be determined.

First check that you can handle this technique of analysis by working with air. Fill your gas burette with water and then, with the subaseal cap removed, run out the water through the tap so that the burette fills with air. Make sure that you are not breathing close to the burette when you do this. Leave about 1 cm³ of water in the burette, close the tap and replace the subaseal cap. Place the burette in a large trough of water and leave it for about five minutes to attain the temperature of the water. Adjust to atmospheric pressure by opening the tap under water, raising or lowering so that the water levels inside and outside the burette are equal, and closing the tap. Record the reading in the burette. Use a syringe to inject about 0.2 cm³ of the potassium hydroxide solution through the subaseal cap. This should be done under the surface of the water. Take great care with the potassium hydroxide solution which is very concentrated and extremely corrosive. If you get any on your fingers, wash them immediately in the bowl of 2 per cent sulphuric acid and then with water. You are strongly recommended to wear an eye shield.

Thoroughly mix the gas in the burette with the potassium hydroxide by tilting the burette to and fro many times. Relevel to atmospheric pressure and record the new volume. Now inject about 0.4 cm³ of the pyrogallol solution and repeat the procedure. Wait several minutes to be sure that the gas is at the same temperature as the water because the reaction is exothermic. The pyrogallol solution contains excess potassium hydroxide. It is very strongly alkaline and it should be treated with exactly the same care as the potassium hydroxide solution itself. Work out the percentage by volume of carbon dioxide and oxygen in the air. Are the values within your expected experimental error? If not discuss the matter with your teacher to see if you can improve your technique.

Now arrange to analyse for carbon dioxide and oxygen the gas remaining in the bell jar after the candle has been burnt. Work out a way of transferring the gas to the burette and discuss it with your teacher. You should also discuss a plan of campaign to find out as much as possible about this apparently simple process. It may be desirable to work in teams. It may be interesting to work with a small vessel, such as a wide glass tube, rather than a bell jar, to see if the greater chance of circulation of all the gases to the candle has a significant effect. In all cases, including the bell jar, you should record the rise in the jar as you will want to relate the change in volume on combustion to the analysis of the product.

Record and Report

Write a full report on your findings. Draw conclusions about the value of the candle experiment for showing the percentage of oxygen in the air and the circumstances which you think govern the extinguishing of the candle.

10: The Use of Sodium Thiosulphate in Volumetric Analysis

10.1 The Stoichiometry of the Iodine–Sodium Thiosulphate Reaction

The reaction between iodine and thiosulphate ion in neutral or weakly acid solution is both quantitative and rapid. The end point can be accurately detected using starch indicator.

Apparatus and Materials

Titration apparatus
0.05 M iodine solution (1 dm³ of solution should contain 12.7 g of iodine and 40 g of potassium iodide)
0.05 M sodium thiosulphate
Starch indicator (make 2 g of starch into a paste with cold water. Add 0.01 g of mercury(II) chloride and boil with 1 dm³ of water; then cool. The mercury(II) chloride is added to prevent the growth of moulds. If the solution does not have to be kept for a long period it can be omitted.)

Procedure

First carry out test-tube experiments to observe the behaviour of the starch indicator. Do not add the indicator initially to the iodine solution, but only after the iodine colour has been reduced to a pale yellow, otherwise iodine becomes adsorbed on to the starch and accuracy in titration is reduced.

Now titrate 20 cm³ portions of 0.05 M iodine solution with 0.05 M sodium thiosulphate until the solution is only pale yellow. Then add 2 cm³ of starch indicator and continue the titration until the indicator colour just disappears. Obtain two accurate titres.

Calculation

The equation for the reaction is

$$n\text{Na}_2\text{S}_2\text{O}_3 + \text{I}_2 \longrightarrow 2\text{NaI} + \text{other product}$$

and therefore

$$n = \frac{\text{moles of Na}_2\text{S}_2\text{O}_3 \text{ in titre}}{\text{moles of I}_2 \text{ in solution}}$$

Find your value for n. Write a balanced equation for the reaction, finding out the other product of the reaction from a reference book. Write an ionic equation for the reaction.

10.2 An Investigation of Bleaches and Disinfectants using Iodimetry

Sodium hypochlorite solution forms the basis of most commercial bleaches. In addition, most contain stabilizers and colouring dyes. In this technique, the sodium hypochlorite reacts with excess of potassium iodide solution, liberating iodine which is then titrated using standard sodium thiosulphate solution. Thus:

$$OCl^- + 2I^- + 2H^+ \longrightarrow I_2 + H_2O + Cl^-$$

and

$$I_2 + 2S_2O_3^{2-} \longrightarrow 2I^- + S_4O_6^{2-}$$

Apparatus and Materials

Titration apparatus
Standard flask (various sizes)
0.10 M sodium thiosulphate
0.5 M potassium iodide
Dilute sulphuric acid

Starch indicator
Commercial bleaches and disinfectants
  containing OCl^-
10 cm^3 burette

Procedure

First use a measuring cylinder to find the total volume of the bleach you have been given. Note down the price of the bleach.

Measure 0.1 cm^3 of the bleach from a small burette into a conical flask. Dilute the sample with 20 cm^3 of water and add 5 cm^3 of 0.5 M potassium iodide, and then 5 cm^3 of 2 M sulphuric acid. Titrate the iodine which is released with 0.1 M sodium thiosulphate, using starch indicator. Use this preliminary experiment to calculate a volume of bleach which can be diluted in a standard flask to give a reasonable titre of sodium thiosulphate.

Run the calculated volume of bleach into a standard flask. Dilute with 20 cm^3 of water and add a mixture of 0.5 M potassium iodide and 2 M sulphuric acid (1:1), until the liberated iodine forms a clear brown solution in excess of potassium iodide. Dilute to the mark with water. Titrate 25 cm^3 portions with 0.1 M sodium thiosulphate.

Calculation

Calculate the molarity of the original bleach used.

Calculate the cost per mole of sodium hypochlorite.

Compare the results with those of others in the class who used different bleaches to choose a 'best buy' in commercial bleach solutions.

11: Investigation of Some Polymers

Polymers make up a class of organic compounds which are of immense importance as they are used as plastics and artificial fibres. In our investigation of them we shall attempt to find out what elements are present in just a few of the wide range of compounds available, and we shall use a technique known as gas–liquid chromatography to discover what are the building blocks from which some polymers are made. When polymers are heated they are often broken down into smaller molecular fragments—a process known as pyrolysis—and chromatography enables us to find out what these fragments are.

We shall first make a simple investigation of the effect of heat on some polymers. This can often give quite useful clues as to the nature of the polymer and the elements present in it. Then we shall do systematic tests for the elements that might be present. We shall not need to test for carbon and hydrogen as these are present in all polymers used industrially. We therefore confine our tests to those for the halogens, nitrogen, and sulphur, using a method which can, in fact, be applied to organic compounds generally. It is known as Middleton's test.

11.1 Effect of Heat on Polymers

Apparatus and Materials

Small test-tubes

Samples of various polymers (poly-styrene, polyethylene, polypropylene, polybutadiene, polyisoprene, nylon, polyvinylchloride, rubber, silk, wool, cotton wool)

Litmus (or universal indicator paper)

Bromine water

Approximately 0.1 M silver nitrate solution

Procedure

Heat 1–2 g of each polymer in a small Bunsen flame in a small, hard-glass test-tube. You may use the polymers available in the laboratory or you can bring in various polymers from home—detergent containers, plastic bags, old stockings or pieces of shirt made of 'man-made' fibres, etc. In your investigation try to find evidence for the presence of any of the following:

(a) acidic or alkaline gases;

(b) unsaturated gaseous compounds;

(c) gases containing halogens.

11.2 Detection of Halogens, Nitrogen, and Sulphur by Middleton's Fusion Test

This test is based on the fact that when organic compounds are heated strongly with a mixture of zinc dust and anhydrous sodium carbonate, any nitrogen they contain is converted into sodium cyanide, any halogens to the sodium halide, and any sulphur to zinc sulphide. The cyanide and halides are soluble in water and their presence in solution is detected by the standard tests. Zinc sulphide is insoluble in water but when it is acidified, it liberates hydrogen sulphide which is readily identified.

As the precipitation of silver halides is used to test for the presence of the halogens, it is necessary to remove any cyanide ion by boiling with an acid. Failure to do this will interfere with the test. Why is this so and why does boiling with an acid remove the cyanide ion but not the halide ions?

Apparatus and Materials

10 × 75 mm hard glass test-tube
Rack of normal sized test-tubes
Filter funnel and papers
100 cm³ beaker
Evaporating basin
Test-tube holder
Zinc reagent made by grinding 25 g of analar zinc dust with 50 g of analar

anhydrous sodium carbonate in a mortar. This must be kept dry.
Iron(II) sulphate
Dilute solution of iron(III) chloride
Approximately 0.1 M silver nitrate solution
Lead acetate solution or lead acetate test papers

Procedure

(a) *Reaction with zinc dust and sodium carbonate.* Place 20 cm³ of distilled water in an evaporating basin in readiness. Cut up the polymer as finely as possible. A coarse file may help. Place about 0.1 g of the polymer in a small hard-glass test-tube and add the zinc dust reagent to a depth of about 1 cm. Shake to ensure thorough mixing and then add more of the reagent to a total depth of 3–4 cm. Hold the tube horizontally in a test-tube holder and gently heat the mixture at the open end of the tube. A micro Bunsen burner will be easier to use. If the mixture is pushed along the tube by escaping gases, rotate it and tap it gently. Gradually increase the temperature to red heat and then slowly move the flame down the tube to the bottom. When no further fumes are given off and the reaction seems to be finished, hold the tube vertically for a moment or two, getting the whole end red hot. Plunge the red hot tube into the water in the evaporating basin, tapping the bottom of the tube to ensure that it breaks completely and the contents come into contact with the water. This part of the procedure is vigorous and requires care. Now boil the water for a few minutes to aid solution. Decant the liquid into a prepared filter funnel and paper. Both the solid left in the basin and the filtrate are required.

(b) *Testing for nitrogen, halogens, and sulphur*

Nitrogen. To a portion of the filtrate add one or two cm³ of dilute sodium hydroxide and then a little solid iron(II) sulphate. Boil for a moment and then cool under the tap. Add one drop of iron(III) chloride solution and acidify with dilute hydrochloric acid with care. Why is a great deal of gas evolved? At this stage a deep blue precipitate will indicate nitrogen. If the colour is faint or even greenish, try filtering. The precipitate may be much more obvious on a filter paper.

Make sure that you know the reaction between iron(II) ions and cyanide ions, and between the product and iron(III) ions.

Halogens. Acidify a portion of the filtrate with dilute nitric acid and place it in a small beaker. Boil for several minutes. Cool under the tap and add a few drops of dilute silver nitrate solution. Carry out a blank test for comparison with distilled water and dilute nitric acid. A white or yellow precipitate will indicate a halogen. To determine which halogen is present, study the solubility of the precipitate in ammonia solution. Silver chloride is soluble in dilute ammonia and silver bromide only in concentrated ammonia. Silver iodide is soluble in neither.

Sulphur. Acidify the solid residue in the evaporating basin with dilute hydrochloric acid. Put a few spots of lead acetate solution on a filter paper and place it above the evaporating basin. Boil the solution. Sulphur will be indicated by a darkening of the lead acetate spots. Try to account for the chemical reactions involved in this test.

Question

If the polymer is strongly heated in a good supply of air and a residue is left, what might it be?

Remove some of the insulating covering from a piece of connecting wire and heat it strongly on a crucible lid or piece of asbestos paper. Is there any residue? What can you conclude about the insulating material?

11.3 Pyrolysis of Polymers and 'Fingerprinting' by Gas–Liquid Chromatography

In this experiment, various polymers are subjected to the action of heat. In general this serves to break up the long polymer chains into small fragments. For example, on heating polyethylene strongly, one of the common fragments might well be ethene, C_2H_4, the original monomer from which the polymer was made.

To study the vapours produced from the heating, which will contain many of the pyrolysis products of low molecular weight, gas–liquid chromatography is used. Gas–liquid chromatography is very similar in principle to adsorption chromatography on a column, or on paper which you have used before (Experiment 8.1). In this case the gaseous sample to be analysed is carried through a column packed with solid material by an unreactive carrier gas—nitrogen.

The column of finely packed solid also contains an adsorbed liquid, and it is the solubility of different components of the gaseous mixture in this stationary liquid solvent which determines the rate at which the different gaseous components pass through the column. Various ways of detecting the different fragments as they emerge from the column may be employed. One of the simplest (and that used here) is the measurement of the changes in electrical resistance caused in a heated platinum wire when the composition of the gas surrounding it changes. This is the result of the change in the thermal conductivity of the gas, which leads to a change in the temperature of the wire. The changes are measured by a Wheatstone bridge circuit which is linked direct to a potentiometric recorder. Such a detector is called a katharometer (Figure 11.1). Figure 11.2 gives a diagram of a gas chromatograph using a katharometer detector which was very simply made by a student, L. J. Hanna, of Chester College.

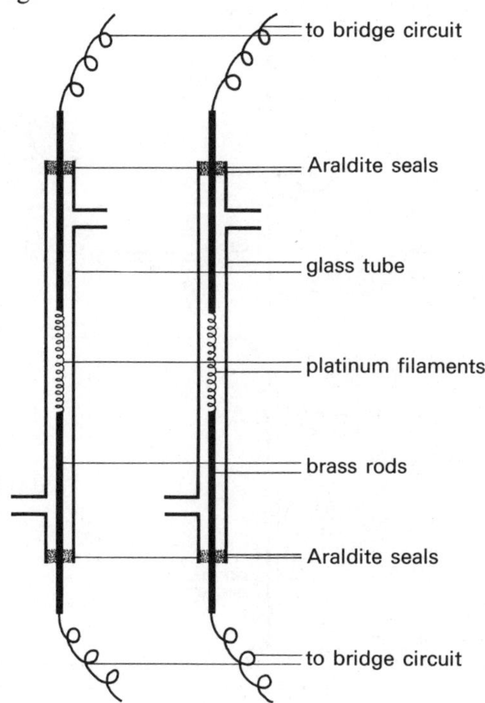

Figure 11.1   The katharometer

Apparatus and Materials

Small hard glass test-tubes
Small U-tube
100 cm³ gas syringe
Samples of polymers

1˙ cm³ hypodermic syringe with fine needle
Apparatus for gas–liquid chromatography (see below for details)

1. Nitrogen carrier gas
2. Injection head with subaseal cap
3. Chromatography column
with stationary phase
4. Needle valve in by-pass tube
5. Platinum wire katharometer
6. Beakers of water to judge
rate of gas flow
7. Recorder or scalamp
8. 4 volt source to heat katharometer
wires and to power the bridge circuit
9. Fixed resistor
10. Variable resistor to control
the balance point

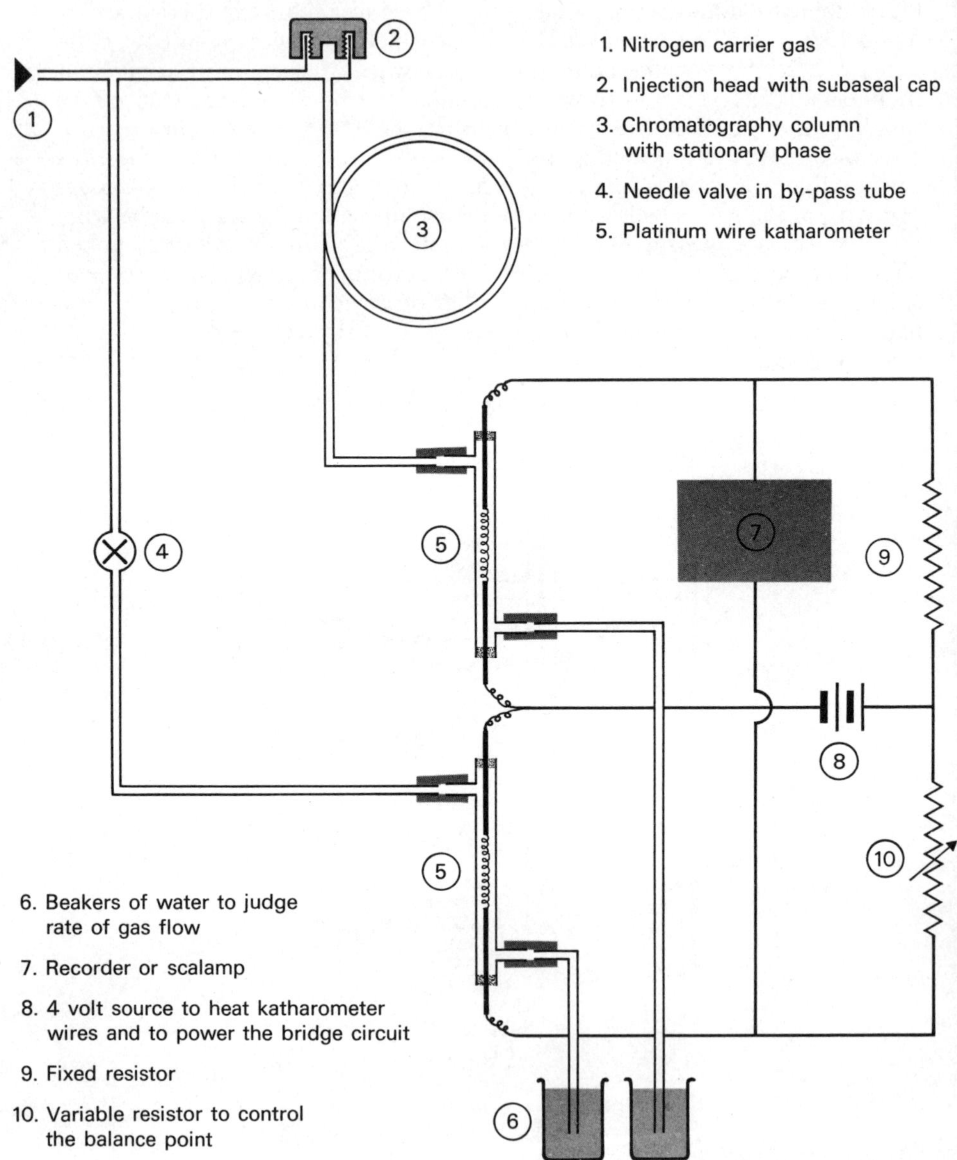

Figure 11.2

Procedure

Pyrolysis samples for examination by gas–liquid chromatography may be easily obtained as follows. Place a small sample of the polymer in a hard-glass test-tube fitted to a small U-tube which stands in a beaker of water and serves to condense any water vapour produced on heating. After initial heating during which air is removed from the apparatus, connect the U-tube to a 100 cm^3 gas syringe which may be used to collect the gaseous products (see Figure 11.3). Material is then available to carry out a number of runs with the gas chromatograph.

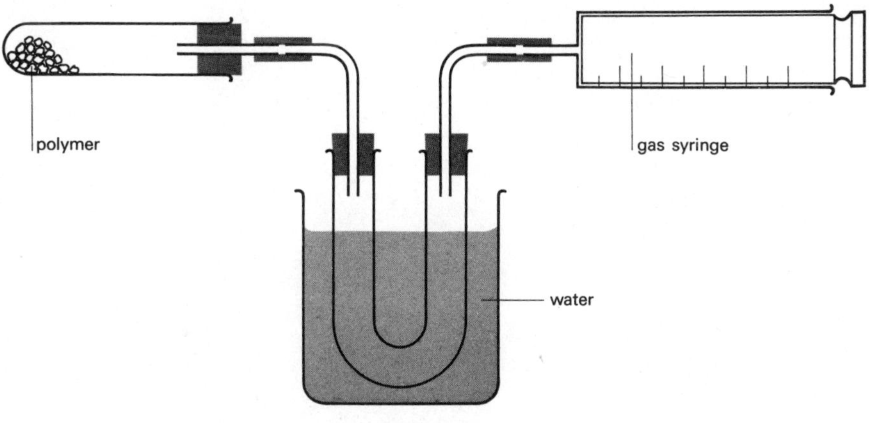

Figure 11.3

Introduction of the sample into the gas chromatograph is achieved by extracting a small sample of the gas from the large gas syringe, using a small hypodermic syringe with a fine needle. Use a glass syringe in preference to a disposable plastic one which will soon stick when exposed to the pyrolysis products. A volume of 0.5–1 cm^3 of the sample will normally be sufficient, but will depend on the gas chromatograph being used. A diagrammatic representation of a gas–liquid chromatograph is shown in Figure 11.2.

It is not intended that a detailed analysis of the traces from the different samples should be made. However, you should study the notable peaks on each trace, and see whether the trace is sufficiently characteristic to allow identification of the polymer from it. It is particularly valuable to compare the traces obtained from pairs of similar polymers, for example, polyethylene and polypropylene, polyisoprene and polybutadiene. All the polymers you study will show a peak in the trace due to the presence of hydrogen. They will all show at least one peak on the other side of the baseline from the hydrogen peak, and this will be due to a hydrocarbon fragment. Study the traces carefully to see whether you can suggest the number of carbon atoms in each of these fragments. Is there any relationship evident between the number of carbon atoms in the fragment and the length of time for elution through the column to occur?

Do you feel that this method gives the basis for a satisfactory scheme for identifying different polymers from their pyrolysis products?

12: Thin-layer and Two-way Chromatography

12.1 Study of the Products of the Nitration of Phenol using Thin-layer Chromatography

In the first part of the experiment phenol is nitrated by reaction with a mixture of sulphuric acid and sodium nitrate. In the second part of the experiment an attempt is made to separate and identify the products of the reaction.

Apparatus and Materials

Test-tubes, 150 × 25 mm
Thermometer (-10 to $110\ ^\circ$C)
Teat pipette
Beakers
Prepared thin-layer plates for chromatography
Melting point tubes
Dilute sulphuric acid

Sodium nitrate
Phenol
o-Nitrophenol (3 per cent solution in ethanol)
p-Nitrophenol (3 per cent solution in ethanol)
Solvent for chromatography (toluene in dioxan, 6:1)

Procedure

(a) First make up the nitrating mixture by dissolving about 2 g of sodium nitrate in 6 cm^3 of dilute sulphuric acid in a test-tube. Cool the solution by standing the test-tube in an ice bath.

In another test-tube melt 1 g of phenol with about 1 cm^3 of water. Add this to the previously prepared solution, a portion at a time, so that the temperature does not exceed 20 °C. (Note: phenol in concentrated solution is caustic!) Shake well after each addition. When all the phenol has been added, allow the tube to stand for one hour, shaking it frequently.

Decant the excess of reaction solution from the resinous product and wash two or three times with small portions of water, discarding the washings. Remove the last traces of water with filter papers. You have made a mixture of nitrophenols and other products.

(b) Thin-layer chromatography may now be used to separate and identify some of the products. The practice of this technique is essentially similar to that of paper chromatography. The great advantage is the very short time taken for the chromatogram to be developed. To save time, thin-layer chromatoplates have been prepared for you ready for use. They consist essentially of a thin layer of silica gel, the particles of which are held together by a binder, usually of gypsum.

Two of the products of the reaction may be *ortho*-nitrophenol and *para*-nitrophenol. You are provided with standard samples of these, dissolved in ethanol. By running these with the reaction product you should be able to gain some information about its composition.

Use melting point tubes drawn out to a fine jet for spotting the plate with samples of *ortho*-nitrophenol, *para*-nitrophenol, and the reaction product (a 3 per cent solution should be made in ethanol). One drop of each of the samples of *ortho*-nitrophenol, *para*-nitrophenol, and the reaction product (a should be at least 1 cm from the edges of the plate and be not larger than 2–3 mm in diameter when dry. The plate should be dried with a hot air dryer after spotting.

Pour sufficient of the toluene/dioxan solvent into a clean beaker just to cover the bottom of the beaker. Place the plate in the beaker, as nearly vertical as possible, and cover the beaker with a watch-glass. Separation should take only a few minutes, after which time remove the plate, mark the solvent front, and measure R_f values for all spots.

Questions

1. How many separate constituents can you detect in your reaction product? Are *ortho*-nitrophenol and *para*-nitrophenol among them?
2. From the relative sizes of the spots, can you estimate the relative proportions of any of the constituents of the reaction products?

12.2 Separation of Amino Acids by Two-way Paper Chromatography and an Analysis of Some of the Amino Acids in Perspiration

As many as twenty-four different amino acids may be present in the hydrolysis products of proteins. The physical properties of amino acids are very alike and the separation of such complicated mixtures is extremely difficult. In 1952 Martin and Synge were jointly awarded the Nobel Prize for Chemistry for work involved in the separation of amino acids. The technique they used was paper chromatography and it allowed the identification of the amino acids in protein hydrolysates. It is the development of this technique which has led to our knowledge of the full structure of many proteins and this in turn has played a major part in modern biological chemistry and in the understanding of the working of the cell.

You have used paper chromatography for the analysis of brass in Experiment 8.1. Now you are going to extend the technique to that of two-way chromatography. The R_f factors for different amino acids may well be very close for any one particular solvent and such a solvent would only separate the amino acids into broad groups. However, a pair of solvents may be chosen such that no amino acids have the same R_f factors with both. If the mixture of amino acids is first run on paper with one of the solvents and then the paper is turned through a right angle before running with the other solvent, each amino acid will move to a different place on the paper (see Figure 12.1). The mixture is applied to the paper as a spot at position X. From the diagram in the figure you will see that solvent A by itself would fail to separate the four acids as acids 1 and 2 have the same R_f factors. Solvent A followed by solvent B at right angles gives a separation of all four acids. Can you see why running two chromatographs with each solvent separately will not allow a complete analysis?

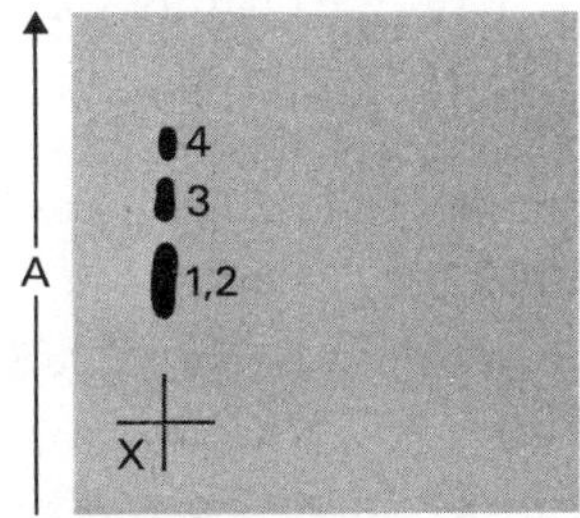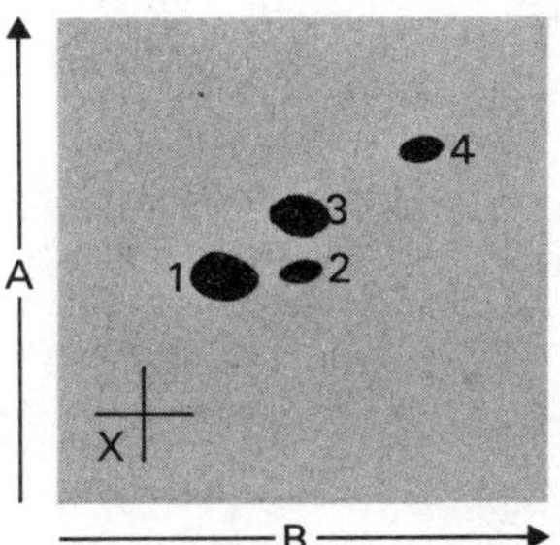

Figure 12.1

The experiment you are going to carry out is restricted to five amino acids: glycine, alanine, lysine, valine, and tryptophan. The first thing is to find out the R_f factors for each amino acid with each of three solvents. From these results you will be able to select a pair of solvents which will allow you to separate all five amino acids from a mixture by means of a two-way technique. Amino acids are themselves colourless but they may be located on the paper by reacting them with ninhydrin with which they give an intense blue-purple coloration.

You should read through Experiment 8.1 in order to check points of technique.

Apparatus and Materials

Whatman chromatography paper CRL/1

3 Chromatography jars ($\frac{1}{2}$ lb Nescafe jars are very suitable and the colour does not matter)

25 × 150 mm test-tubes with corks. These are only required if the strip technique is used.

Number 2 cork borer

Scissors

Melting point tubes

Clips to hold the paper in a cylinder (see text for details)

Hot air dryer

Spray bottle or aerosol for ninhydrin

0.1 per cent solutions of amino acids in a solvent made of one part of propan-2-ol and nine parts water

Amino acids: glycine, alanine, valine, lysine, and tryptophan

Solvents:

(i) 80 parts ethanol (colourless methylated spirit will do), 20 parts water

(ii) 60 parts butan-1-ol, 15 parts glacial ethanoic (acetic) acid, 25 parts water

(iii) 60 parts butan-1-ol, 20 parts 2 M ammonia, 20 parts water

Locating agent: 0.2 g of ninhydrin in 100 cm^3 of acetone. This must be kept in a refrigerator or it will deteriorate quite quickly.

Procedure

(a) Two techniques are described. The first one is suitable for a fairly large group of people, each doing a single run of one amino acid with one solvent. The results may then be pooled before going on to the next part of the experiment using the two-way technique. The three solvents are:

(i) 80 parts ethanol, 20 parts water
(ii) 60 parts butan-1-ol, 15 parts glacial ethanoic acid, 25 parts water
(iii) 60 parts butan-1-ol, 20 parts 2 M ammonia, 20 parts water

Perspiration contains amino acids. To avoid stains on the paper it is essential to wear rubber gloves throughout the experiment.

Technique 1. Place one of the solvents to a depth of about 1 cm in a 150 × 25 mm test-tube. Take care not to get solvent on the walls of the test-tube by using a teat pipette. Cork the tube and leave it while you prepare the paper. What is the purpose of doing this? Cut a strip of chromatography paper, 1 cm wide and 14 cm long. Mark a start line with a pencil (not a ball-point pen) 2 cm from one end. Score a line down the centre of the strip of paper with the back of the blade of a pair of scissors, so that it can be easily folded to give an angular cross-section. This gives the paper rigidity and stops it flopping around in the test-tube. The paper should not be folded at the end with the start line. Use a number 2 cork borer to make a hole part way through the narrow end of the cork. Suspend the strip of paper from the cork by inserting it in this hole. Use a finely drawn out melting point tube to apply spots of an amino acid solution to the centre of the start line. Apply about three spots. Dry with a hot air dryer after the application of each spot so that the final spot is not more than 3 mm diameter. Place the strip in the tube, ensuring that the end is beneath the surface of the solvent and that it is not touching the walls and allow it to run until the solvent front is close to the top of the paper. This takes about 1½ hours. Mark the solvent front with a pencil and dry the paper. Proceed to Technique 3 below.

Technique 2. This is an alternative technique when there are fewer people involved in the experiment. Place solvent to the depth of about 1 cm in the jar you are going to use and close the jar. Cut a piece of chromatography paper 15 cm square and mark a start line with a pencil 2 cm from one end. Mark five points on the start line at 2 cm intervals and not closer than 3 cm to an edge of the paper and label these points with the names of the five amino acids. Place spots of each of the amino acids at the five points, using a finely drawn out melting point tube. Three spots should be sufficient for each and they should be dried with a hot air dryer between each application. No spot must spread to a diameter greater than 3 mm. Roll the paper into a cylinder and fix it with suitable clips. Place it in the solvent and allow it to run until the solvent is very close to the top of the paper. Mark the solvent front and dry the paper.

Technique 3. Spray the paper in a fume cupboard with the ninhydrin solution and place the paper in the oven at 105 °C for about five minutes. Mark round the developed spot with a pencil and record the R_f factors for each amino acid. Each spot will spread a little way and you should record the R_f factor as being between two limits.

From these results select the pair of solvents for your two-way run. Take a piece of chromatography paper 15 cm square and mark a start point 2 cm in from each edge at one corner. Apply between five and ten spots of the mixture of amino acids. Use the two-way technique with the two solvents, drying the paper between the two. Locate the spots as before and hence identify the amino acids which are present in the mixture.

(b) *The investigation of perspiration*. A mixture of 10 per cent propan-2-ol in water is a good solvent for amino acids. Think of a good way of getting plenty of perspiration on some part of your skin; a rubber glove may help. Holding a small piece of filter paper, about 4 mm square, in a pair of tweezers, mop up the perspiration. Dissolve this in a few drops of the propanol solvent in a very small test-tube. The object is to make the solution of the perspiration as concentrated as possible. Apply the technique of two-way chromatography, which you have just used, to separate the amino acids. Can you say whether any of the five amino acids you used in the first part of the experiment are present in the perspiration? State very clearly what you have found out about the amino acids in perspiration.

13: Using the Flow of a Bubble in Solution

13.1 Determination of the Concentration of an Aqueous Solution of Ethanoic Acid using a Method based on the Flow of a Bubble in the Solution

In this book you will have experienced a number of methods of quantitative analysis that depend on some property of a material changing with a change of composition. This property may have been such things as colour, electrical potential, and so on. The purpose of this experiment is to allow you to investigate a property which you have never met before and then to see if it can be used for analytical purposes. As a result you should appreciate the basic principle of many analytical methods; that is, the setting up of a consistent set of data for known mixtures and finding out how the data for an unknown mixture relate to this.

The property we are concerned with is the flow of a large bubble of air through a tube containing a liquid. The rate is dependent on the properties of the liquid in the tube and on the diameter of the tube but it is independent of the length of the bubble provided that the bubble is at least as long as three tube diameters.

You are to investigate whether the measurement of the rate of flow of an air bubble through a solution of ethanoic acid in water will enable you to arrive at the composition of the solution.

Apparatus and Materials

A glass tube of internal diameter about 0.7 cm and length 1.5 m, fitted with corks at both ends
Stop-watch
Retort stand and clamp
100 cm^3 graduated cylinder

Burette
6 × 100 cm^3 conical flasks
Small funnel to aid in pouring liquids into the long tube
Glacial ethanoic acid

Procedure

For this experiment you will use a glass tube about 1.5 m long. Stick two pieces of sticky paper to the tube so that you can mark two points exactly 50 cm apart over the central section of the tube. Cork one end of the tube and fill it with distilled water to within about 5 cm of the top. Cork the top end and clamp the tube at its middle in such a way that it is vertical but can easily be swung through 180 degrees until it is vertical in the opposite position. This will involve loosening the clamp, swinging the tube, checking that it is vertical and reclamping the tube. Time the rise of an air bubble over the 50 cm distance and take sufficient readings to be satisfied with the consistency of the result.

Now investigate the rates of flow of a bubble of similar size in the same tube with solutions of ethanoic acid in water. Keep to a range of concentrations between 0 and 10 per cent of ethanoic acid. Decide your own method of making up the solutions. If they are within ± 0.2 per cent they will be accurate enough. Choose a sufficient number of solutions to allow you to report on the feasibility of determining the percentage by volume of an unknown solution of ethanoic acid within this range of concentrations. If you find that the method is feasible you should test it by getting someone to make up a solution within this range for you to analyse.

Record of your Work

Write a report on your investigations, giving the precise details of the techniques used and the results obtained. State clearly whether you consider this a suitable method of analysis and give the limits of the accuracy that can be expected.